1

THE THINGS THAT ARE WEIGHING YOU DOWN: A LOOK INTO MENTAL ILLNESS

There was a time when our population had to deal with things like civil war, violent crime, the black plague, violent crime, and the struggle for food and water. These were stressful times. That level of stress, commonly referred to as acute stress, was tough. Perhaps our ancestors often died from stress than any other disease but just didn't know it.

Unfortunately, stress is still a problem today, even though we are more developed and civilized. Why? Because although we're not dealing with war and other external threats, there continues to be an increase of low-level chronic stress that's silent and hidden. A person gets up in the morning, drowns themselves in coffee, and reports to an office job where they sit in their cubicle from 9am to 5pm dragging themselves from one task to another and periodically faking a smile at the watering hole. To all their colleagues, they're just aloof, absent-minded, sometimes moody, and often slow to complete projects.

Unknown to them is that this person is fighting a hidden battle that no one else knows about. Every morning is a drag. The sun never seems to shine, and there seems to be a grey cloud looming over them wherever they go.

This is a common experience for many suffering from low-level chronic stress. Yet, there are many more symptoms that could vary from individual to individual. The natural process of the body's stress response is to trigger heightened alertness and energy for a short period as the fight, freeze, or flee response is activated. When this becomes a regular everyday occurrence, devastating consequences follow. But here's what I want you to understand. Stress doesn't necessarily lead to mental illness unless left unmanaged for prolonged periods. And if you are struggling with mental illness, there's nothing wrong with you.

WEIGHED DOWN BUT NOT ALONE

Contrary to what the voices in your head might be telling you, you're not crazy or making things up. Something is wrong, and you do need to heal from this current state. However, you shouldn't feel shame, remorse, or assume that you're weak for falling mentally ill.

In fact, you're not the only one struggling with mental illness. Take a look at some of these statistics.

970 million people worldwide have a mental health or substance abuse disorder, according to Our World in Data. The same report issued in 2018 showed that anxiety is the most common mental illness in the world. Pre-pandemic (COVID-19, which impacted all of us in

2020), there were 248 million reported cases of people suffering from this mental illness. Females are more affected by mental illness (11.9%) than males (9.3%). JAMA Psychiatry also reported an alarming statistic back in 2015 when they said that an estimated 14.3% of deaths worldwide are attributed to mental disorders. As if that's not disheartening enough, here's what we know from 2020.

- More than 264 million people suffer from depression globally. (source: World Health Organization).
- Depression is the leading cause of disability in the world. (source: World Health Organization).
- 17.3 million adults in America (that is, 7.1% of the adult population) have had at least one major depressive episode.
- Adolescents between 12 to 17 years of age had the highest rate of major depressive episodes (14.4%), followed by young adults between 18 to 25 years of age (13.8%). 11.5 million adults also had a major depressive episode. (source: Substance Abuse and Mental Health Services Association).
- Anxiety is affecting approximately 40million adults in the United States each year, according to the Anxiety and Depression Association of America.

Given the dramatic global pandemic that devastated all of us, I anticipate these numbers will continue to rise. I'm sharing these painful statistics with you for one reason. I need you to realize that you're not the only one struggling.

There are millions of us across the globe silently struggling with the difficulties of life. Some are fortunate enough to make it through the dark storms of life better than others. When crises or challenges show up, these individuals seem to have the power to transform into their own superhero and face the obstacle head-on. Ultimately, they emerge out of the difficulty more resilient and victorious. But not everyone comes in-built with this natural ability. Some of us get hit, and we lack the strength or mental power to get back up.

I am writing this book for you if you can feel in your heart that you're ready to follow simple, proven techniques to heal yourself and conquer all mental disorders from your life. If what you want is a new chapter of your life to begin where you are stronger, healthier, happier, more confident, and at peace, this book can become your guide.

It is not written with sophisticated medical jargon that makes me sound impressive and smart but yields no results for you. I've deliberately simplified every technique, concept, and suggestion so that anyone from any background can understand and apply these teachings.

One thing I must mention before going any further is that you cannot use this book to self-diagnose or as a substitute for professional medical help. It is not intended for that, so always make sure to consult a qualified medical professional, especially if your case is extreme.

ANXIETY, DEPRESSION, OCD, INTRUSIVE THOUGHTS: WHAT ARE THEY?

Mental illness, also called mental health disorders, refers to a wide range of mental health conditions, the most common of which are anxiety, depression, OCD, and intrusive thoughts.

People will experience mental health concerns at different stages of their lives, especially when there has been prolonged stress. And should those concerns persist, i.e., continued and uncontrollable mood disorders, eating disorders, etc., this can quickly turn into a mental illness. Let's touch on the most common ones.

ANXIETY

Anxiety is usually a natural reaction of the body when under stress. It's a strong feeling of fear about what's to come. Think of it like worry on steroids. It's expected that we experience some anxiety when doing something like a job interview, giving a speech, a big final exam, or anything else that triggers your nerves and causes you to be fearful. What causes anxiety? We don't really know, but researchers are still working on an answer. It's probably a combination of genetic disposition, lifestyle, brain chemistry, and other environmental factors.

There's a great story of the famous Opera Singer Caruso that's often told. He got nervous, and anxiety kicked in just before one of his big performance. Minutes to curtain raise his throat dried up, he couldn't sing, and he experienced stage fright. Thankfully, he was aware of how to handle his anxiety and calm his nerves. That single performance put his name on the map in the world of Opera, and the rest, as they say, is history.

So, anxiety is not so much a problem as it is a defense mechanism. However, what goes wrong is that these extreme feelings don't just last a few minutes for some of us. They go on for months on end. Daily functioning at work and at home becomes a problem.

There are many types of anxiety disorders, including panic disorders, phobia, social anxiety disorders, obsessive-compulsive disorders, separation anxiety, illness anxiety disorder, and post-traumatic stress disorder.

- **An anxiety attack** - is a feeling of overwhelming apprehension, worry, distress, or fear. Different people will experience this in diverse ways. Some might feel dizzy, sweat profusely, get hot chills or hot flashes, become restless, etc. Others might get dry in the mouth, become distressed, experience shortness of breath, and so much more.
- **Panic disorder** - occurs when you experience panic attacks at unexpected times. Most people who suffer from panic attacks live in constant fear of the next episode.
- **Obsessive-compulsive disorder** - recurring irrational thoughts that lead you to perform specific, repeated behavior.
- **Phobias** - these are due to extreme fear of a specific activity, situation, or object.

- **Social anxiety** - occurs when one fears being judged by others in social situations.
- **Separation anxiety** - is fear of being away from loved ones or extreme homesickness.
- **Illness anxiety disorder** - occurs when you get too anxious about your health.
- **Post-traumatic stress disorder (PTSD**) - is anxiety following a traumatic event.

Depending on your personality, temperament, and environment, anxiety will take many forms. It can be as easy to spot as Caruso's story of getting a dry throat and being unable to activate his voice, or it can be in the form of butterflies in your stomach or a racing heart. It can also feel overpowering and out of control to the point where you have trouble breathing, and you feel a disconnect with your mind and body. Some people experience repeated nightmares, painful thoughts, or memories that they can't seem to shut out or control. But typical symptoms that we see across the board is a deep feeling of constant fear and worry. You might also have noticed rapid breathing, increased heart rate, trouble concentrating, difficulty falling asleep, restlessness, among a multitude of other discomforts.

How to diagnose anxiety

A single test will likely not diagnose anxiety. So, you will need to see a professional and go through a lengthy physical examination process, mental health evaluations, and psychological questionnaires. Some doctors might even put you through blood and urine tests to rule out underlying medical conditions.

Treatments for anxiety

Depending on how severe your anxiety disorder is, you may need medical treatment, or you might be able to self-heal with some lifestyle and mindset changes. Officially, there are two categories of treatment: medication and psychotherapy. Unless you're dealing with something extremely severe, I ask that you also consider a third alternative - natural remedies and self-healing. The central aspect of using this third method is to increase your awareness and understanding of your mind and body. It's about learning to take better care of yourself and your mind. You'll notice that although this book teaches lots of therapy-based techniques, I will continue to encourage you to eliminate unhealthy habits and develop healthy mind-body practices.

DEPRESSION

Most research points out that depression affects more women than men. Still, it could be because most men don't immediately seek professional help when they realize something is wrong. Although depression is high in teens and young adults, it can strike at any age for various reasons. Since the global pandemic of 2020, where many people had to struggle with financial uncertainty, job loss, and social distancing, causing them to see loved ones less, it's probably triggered an increase in depression for adults in their 30s and 40s. According to the Centers for Disease Control and Prevention (CDC), adults in the United States reported elevated levels of adverse mental health conditions, substance use, and suicidal ideation in the United States in June 2020. The prevalence of anxiety disorder symptoms was approximately three times those reported in the second quarter of 2019

(25.5% versus 8.1%). The prevalence of depressive disorder was about four times what was reported in the second quarter of 2019 (24.3% versus 6.5%).

But what is depression?

It is a mood disorder that causes a persistent feeling of sadness that just eats away at your heart, health, happiness, and productivity. You can't shake this deep sense of loss no matter what you do. It affects how you think, feel, behave, and leads to various emotional and physical problems. Most people suffering from depression can barely get out of bed, let alone carry out normal day-to-day activities. Contrary to what others might think, depression isn't "holiday blues," and it's not a weakness you can just snap out of. For most people, treating depression happens over a long period of time. It often requires both medicine and some form of psychotherapy.

Symptoms of depression:

- Overwhelming sense of sadness, tearfulness, emptiness, and hopelessness.
- Angry outbursts, irritability, and frustration even over trivial matters.
- Slowed thinking, speaking, and body movements.
- Inhibited thinking and creativity.
- Difficulty concentrating, memory loss, and poor decision-making skills.
- Sleep disturbances, including insomnia or sleeping too much.
- Serious fatigue and lack of energy and enthusiasm to do anything. Small tasks require a lot of extra effort.

- Unexplained physical problems such as back pain or headaches.
- No interest or pleasure in most or all normal activities such as socializing, sex, hobbies, physical exercise, etc.
- Feeble appetite and weight loss, or for some, it's the opposite with increased food cravings and weight gain.
- Anxiety, agitation, or restlessness.
- Feelings of worthlessness or guilt.
- Self-loathing, fixation on past mistakes and failures, and a lot of self-blame.
- Frequent or recurrent thoughts of death, suicidal thoughts, a suicide attempt, or suicide.

Did any or all of these symptoms trigger something in you? Have you noticed that you're having noticeable problems handling your day-to-day activities such as school, work, relationships with others, etc.? Perhaps you've been feeling like a dark grey cloud looms over your head wherever you go, and it's not going away no matter what you try. If so, it's time to get some help because you might be suffering from depression.

It's good to recall that your experience of depression may not be as obvious as what you read online because depending on your gender, age and environment, it may show up different. For instance, men usually experience symptoms related to their mood such as anger, restlessness, irritability or they may become too aggressive. It could also be a feeling of emptiness or a deep sadness. Sometimes it can be a rapid shift in behavior e.g., no longer finding pleasure in their favorite activities or losing sexual interest. Physically it can be digestive problems, fatigue, pains and headaches. For women, it's mainly a feeling of hopelessness, thoughts of suicide, thinking or talking more slowly, changes in appetite, increased menstrual cramps and sleep problems.

Types of depression

Atypical features - This type of depression includes the ability to temporarily become cheerful when outer circumstances such as a happy event occur. It also comes with increased appetite, excessive need for sleep, sensitivity to rejection, and a heavy feeling in the arms or legs.

- **Melancholic features** - This type of depression is quite severe. You hate waking up in the morning, struggle with feelings of guilt, agitation, and sluggishness. Your appetite also changes drastically.
- **Anxious distress** - This type of depression causes you to be unusually restless with constant worry about possible events or loss of control.
- **Peripartum onset** - This type of depression occurs during pregnancy or in the weeks or months after delivery, in which case we call it postpartum.
- **Mixed features** - Simultaneous depression and mania, including elevated self-esteem, talking too much, and increased energy.
- **Psychotic features** - This type of depression is accompanied by delusions or hallucinations, which may involve personal inadequacy or other negative themes.

- **Seasonal pattern** - This type of depression relates to changes in seasons and reduced exposure to sunlight.
- **Catatonia** - This type of depression includes motor activity that involves either uncontrollable and purposeless movement or fixed and inflexible posture.
- **Premenstrual dysphoric disorder** - This is a condition that causes depressive symptoms. It's due to hormone changes that begin a week before and improve within a few days after the onset of your period. They should become minimum or completely gone after the completion of your period.
- **Disruptive mood dysregulation disorder** - This condition commonly occurs in children. It often leads to depressive disorder or anxiety disorder during their teen years or adulthood and includes chronic or severe irritability and anger with frequent extreme temper outbursts.
- **Bipolar I and II disorders** - These are mood disorders that are usually hard to distinguish from depression itself. They both include mood swings that range from highs (mania) to lows (depression).

If you're reading this as a parent of a young child or teen trying to identify whether or not they might be suffering from a depression disorder, look out for the symptoms already mentioned and realize that depending on your child's age, there could be other subtle warning signs.

For example, younger children tend to become clingy, refusing to go to school or leave the house. They might get extremely irritable. You might also notice they complain of constant aches and pains, they seem to have a worried look all the time, and their weight might drop drastically. In teenage children, their body language speaks volumes. You might notice your child is extremely negative, always self-loathing, carrying around a sad, defeated look all the time. Their grades often drop due to poor performance and low attendance. You might notice getting to school is a chore. Even if they go to school, they find ways to ditch class, hang out with the wrong kids, start using recreational drugs or alcohol, and their sleeping and eating habits will become anything but healthy. They might also avoid social interaction or even try to get out of family time with you.

Should you see a doctor when you recognize these symptoms?

If you feel depressed, it's a good idea to make an appointment to see a family doctor or a mental health professional as soon as possible. But if you don't like seeking medical treatment, then reach out to a trusted loved one, a friend, your faith leader, or something respectable that you trust. Going through this book and implementing the ideas discussed will also help you heal and get back to your truly happy and energetic state. But I will insist that you seek emergency help by calling your local emergency number if you feel you're at a point where you may hurt yourself or attempt suicide.

The cause of depression isn't entirely known, as is the case for almost all mental disorders. It's always going to be various factors, including brain chemistry, changes in your body's balance of hormones, genetics, and other biological differences. Research is still ongoing to determine exactly what causes this mental disorder. Regardless, I wouldn't fret too much about the cause. Instead, I would be more focused on the journey of healing from depression before it completely debilitates your life.

Do you know your triggers?

Research has shown certain factors increase the risk of getting into major depression. These include a traumatic or stressful event such as the death of a loved one, financial problems, sexual abuse, etc. It could also be triggered by abuse of alcohol or recreational drugs. Suppose you suffer a chronic illness like heart disease or stroke. In that case, depression might get triggered, especially if you are predisposed to get it due to a family medical history of depression. If you have a history of other mental disorders such as eating disorders, anxiety, or PTSD, you could also trigger depression. All this to say, it's crucial to understand self-care and to monitor yourself when faced with triggers so you can mitigate their influence.

Depression can also influence some chronic health conditions making it worse. It is considered a serious medical condition and even especially when one is already fighting a chronic illness such as cancer, cardiovascular disease, asthma, diabetes, or arthritis. If you notice something is off with your emotional state and cognitive functions for an extended period of time, don't just assume it's the blues. Ask for a professional diagnosis to avoid further medical complications as you work on your chronic illness.

Treatment options for depression

If you have severe depression, you may need a hospital stay or an outpatient treatment program until symptoms improve. However, for most people, healing depression requires a combination of medication, some form of therapy, and a lot of self-care. When it comes to medicine, don't wing this. Speak to your doctor and learn about the different options so you can figure out which antidepressant will be right for your case. You might also need psychotherapy, and this book will help you understand some of the best therapies for treating depression. Besides that, you will need to make some changes to your current lifestyle, and we'll be talking more about this in upcoming chapters. Recently, doctors have started recommending other procedures known as brain stimulation therapies such as electroconvulsive therapy (ECT) or Transcranial magnetic stimulation (TMS), both of which are expensive.

OCD

Obsessive-compulsive disorder (OCD) has become a ubiquitous term. It's considered a long-lasting condition that can develop in childhood and worsen or dampen with time, depending on various factors. What is OCD? Well, to answer that, let me first paint a common scenario.

Have you ever left your home to go to a friend's party only to find yourself wondering whether you turned off the cooker and locked the door? For the average person, this thought might come, and in a little while, he or she will conclude that all is well at home and move on to enjoy the party. For someone with OCD, it would become an obsessive thought that keeps replaying over and over. Behavior and mood would be impacted as a result, and it would probably ruin the whole party experience. This is what OCD is. It's an obsession and compulsive behavior.

Depending on how mild or severe your OCD is, you're likely to portray certain symptoms. Ongoing research is starting to point to genetic disposition, environment, and particular brain structure and functioning as the factors that lead to developing this mental disorder. Some individuals with OCD also develop a tic disorder. These are sudden, brief, repetitive moments that include eye blinking, facial grimacing, shoulder shrugging, throat clearing, sniffing, or grunting sounds, among a multitude of other signs.

Treatment options

Typically, your doctor will prescribe mediation, psychotherapy, or a combination of the two if you're diagnosed with OCD. The techniques we will learn in this book may also be beneficial in your journey to healing.

Intrusive Thoughts

What are intrusive thoughts? These are thoughts that seemingly come from nowhere and set up camp in your mind. No matter how hard you try to chase them away, they don't budge. They frequently reoccur without your volition and usually create anxiety within you because they are negative, violent, disturbing, and out of integrity with your true self. Anyone can experience intrusive thoughts (in fact, we all do from time to time). Reported cases of patients range over 6 million in the United States alone, and those are just the few who have enough courage to ask for professional help. Although having these intrusive thoughts doesn't automatically mean you need medical attention, it can be an excellent way to determine whether you have a developing mental health condition. Unsolicited ideas of violence or inappropriate sexual fantasies that keep recurring aren't signs of a healthy mind. So, if you've been dealing with any thoughts that interfere with your mood, character, and daily activities, it might be a good idea to speak with a mental health professional.

When those intrusive thoughts become uncontrollable, they turn into obsessions, which may lead to compulsions. A great example of something small that can grow is the story I shared of worrying about locking your door and turning off your cooker. Suppose you realize that each time you leave the house for something important like a job interview, a friends' party, etc., you struggle to be present because of the constant worry and thoughts such as "did I lock the door?" or "I think I forgot to turn off the cooker." These thoughts aren't just passing warning signals; they literally set camp in your mind and hijack your attention, making it almost impossible for you to remain calm and focused in the present. In that case, you might want to monitor yourself more to figure out if perhaps you're suffering from intrusive thoughts.

The cause of intrusive thoughts in some individuals is an underlying mental health condition such as PTSD or OCD. It could also be due to brain injury, Parkinson's disease, or dementia. I want to invite you now to notice how disruptive your thought patterns have become. Do you frequently have obsessive thoughts? Are these thoughts seemingly glued to your mind? Are they thoughts of disturbing imagery?

Treatment for Intrusive thoughts

Don't feel ashamed if you realize you suffer from this. The best thing to do is to reduce your sensitivity to the thought and its contents. That's where the techniques you'll learn throughout this book become highly valuable. Through Cognitive Behavioral Therapy (CBT), you can relieve and ultimately heal this condition. It's more like talk therapy, where you'll learn new ways of thinking and reacting so you can become less susceptible to intrusive thoughts.

Another critical thing is to focus on self-care. You can create healthy coping strategies and manage your stress better. By recognizing that these intrusive thoughts are just thoughts and that you don't need to label them, you already begin the soothing process that will weaken their grip. You should also figure out the simple things you can do to manage your stress levels so you can bring your mind to a sense of calm.

However, if your condition is pretty bad, I suggest speaking with your doctor so they can recommend medication or a therapist to discuss these thoughts.

WHY DO I HAVE IT?

It took me a long time to finally own responsibility for my mental health. For most of my life, I knew I had an illness that haunted my life ever since I was a child, but I hadn't yet acknowledged what it was or that I had the power to heal. As a young child, I had a loving family, and all my needs were met. I went to the best schools in my hometown, and many could argue I had an ideal childhood. But here's the thing. I didn't feel that way. There was a nagging feeling like something was wrong with me that I just couldn't shake. I have flashbacks of being six years old and probably even before that, and I felt like an outsider looking in when I considered my peers. Things bothered me that didn't bother anyone else. I felt like I didn't belong, and I was different in a way that wasn't loveable. I constantly worried and tried to get as much validation from my parents as I could. But always felt like I was failing them. Depression first kicked in as I became a teen, but long before that, I would have meltdowns, anxiety attacks, and struggled with eating disorders pretty often. But somehow, I managed to grow up with these recurring cycles and fell in love. That was when things went south, and I fell into the worst depression I had ever experienced.

After recognizing that I had a mental disorder that I needed to heal, the question that kept lingering in my mind was, "why me?" I was already having a hard-enough time accepting that my six-year relationship with the woman I intended to marry had just gone up in flames.

Consumed by my rage and her betrayal, I felt the ground beneath my feet open up and swallow me alive. One moment I was the happiest man in the world with dreams to accomplish, and the next, I couldn't eat, sleep, and quite frankly, life lost all meaning. My fiancé didn't just break my heart with her betrayal. She left me feeling hopeless and lifeless. I could not carry out daily activities, and a few weeks after our breakup, I was in major depression.

What I didn't realize at the time is that asking the question "why me?" does nothing productive for your brain or personal recovery. The question is rooted in victimhood (pretty much how I felt all the time). When mental health problems arise, the first thing we want to do is acknowledge, embrace, and find simple ways of empowering and soothing ourselves. I was doing the opposite, and I think many of us are guilty of it too.

But, if you're wondering why you're struggling with mental health issues or why you keep falling into depression, it's good to know that mental illness isn't caused by any particular thing. Several factors contribute, including your family genes, family health history, personal life experiences (childhood abuse or trauma), chemical imbalances in the brain, traumatic brain injury, having a severe medical condition like cancer or AIDS, loneliness, and a deep sense of isolation, etc. Most of the time, it's a combination. In my case, it was the combination of family health history, losing my job, and catching my fiancée cheating on me. Although scientists would like to find a one-size-fits-all root cause, this isn't yet possible. Certain disorders, such as schizophrenia and bipolar disorder, fit the biological model. However, other conditions, such as depression and anxiety, do not.

The brain is extremely complex and more powerful than any super-computer known to mankind, so it's naive to assume every mental disorder can fit into one category. The journey to having proven scientific causes for every mental illness is at infancy, so we best give scientists time to help us accurately answer this question. For now, we need to focus on doing the best we can with the available knowledge.

The other thing I want you to know is that you're not alone in this fight for your life back. You're also not the only one struggling with mental disorders. More than half of Americans will be diagnosed with a mental illness at some point in their life, especially as our society continues to change. I suspect this will be true of almost any developed country you can name. As economies become more volatile, life becomes more fast-paced, and our relationship dynamics continue to evolve, we will likely see more and more people struggling to cope with all these changes and experiences.

If things seem to be going wrong in your life and your mind feels out of control, it's advised that you take steps to see a medical professional who can diagnose what's ailing you. Usually, the diagnosis is pretty simple, and the earlier you catch your condition, the better. It will involve getting to know your medical history, doing some physical examinations, and taking some lab tests depending on your symptoms. You will also take a psychological evaluation where you'll be asked to share your thoughts, feelings, and how you've been reacting lately.

WHAT TO NOTE: THE SIGNS AND SYMPTOMS

While every mental disorder carries its own set of symptoms, some commonalities act as initial red flags. These are the warning signs you need to look out for if you suspect something is terribly wrong, whether or not you've received a diagnosis from a medical professional.

- Drastic sleep or appetite changes.
- Mood changes.
- Significant drop in productivity and ability to function well at school, work, or social activities such as physical exercise.
- Difficulty thinking.
- Apathy.
- Withdrawal and loss of interest in activities previously enjoyed.
- Feeling disconnected from oneself and one's surroundings.
- Nervousness, constant fear that doesn't go away.
- Difficulty focusing and concentrating.
- Unusual behavior.
- Substance abuse.
- Feeling guilty or worthless.
- Constant heaviness and sadness.
- Extreme worry that also brings about physical effects such as shortness of breath, recurring headaches, heart palpitations, restlessness, or a racing mind.

A BATTLE THAT CAN BE WON

It's important to know that regardless of your specific mental health problem, the options available for treatment are numerous. You should consult with a qualified professional on your best options depending on how chronic your case might be. In some instances, medication is necessary and must be part of your treatment. However, therapy can be enough, especially if you feel self-empowered enough. This book focuses on healing through therapy and self-care techniques. We tackle Cognitive Behavioral Therapy (CBT), Dialectal Behavioral Therapy (DBT), and Acceptance Commitment Therapy (ACT). By the end of it, you will know what they are, how they work, and how to apply them now.

2

PRINCIPLE BEHIND COGNITIVE BEHAVIORAL THERAPY (CBT) AND MORE

Cognitive Behavioral Therapy or CBT helps people understand how their patterns of thinking influence their actions and feelings. It's a form of therapy that takes into account how behavior impacts thoughts and feelings.

Many experts in the world of psychotherapy consider this form of therapy the industry's gold standard mainly because it's proven very effective. It's also the most researched form of psychotherapy. It aligns with most international guidelines for psychological treatments making it the first-line treatment for many disorders.

WHY HAS CBT BECOME SO POPULAR?

Many consider CBT the new face of psychology because, unlike the traditional form of therapy that most of us hear about (where a patient spends years lying on the couch in a therapist's office passively trying to get to the root of his or her problem), this new form of treatment is more proactive. The patient and therapist have to work together to develop solutions to the issues at hand. What I like best about this approach is that it's focused on moving forward, creating healthier patterns, and learning to let go of unhealthy ones instead of focusing on events from the past.

Most patients who have experienced this treatment report that they loved the feeling of being empowered and working as a team with their therapist. It's very liberating to feel like you have control over your life. After all, a key aspect of mental health issues is the racing mind and sense of loss over one's power. Even for those dealing with chronic cases where medication is required, this form of therapy still works. It may also help when:

- Coping with a severe medical illness.
- You need to prevent a relapse of mental illness symptoms.
- Coping with grief or loss.
- You want reliable, simple techniques for coping with stressful situations such as job loss, global pandemic, relationship conflicts, etc.

HISTORY OF CBT:

This form of psychotherapy is founded on Albert Ellis and Aaron Beck's work back in the 1950s. Since then, other psychologists and psychiatrists have put together their own techniques and treatment programs based on these ideas. Even DBT (you'll learn about that shortly) evolved from CBT.

In 1955 Albert Ellis proposed his ABC model founded on his belief that external events don't automatically trigger negative emotional responses. What matters is the belief one has about that event. ABC is an acronym for **A**ctivating events, **B**eliefs, and **C**onsequences.

A simple way to think about this is that our emotions and behaviors, i.e., **C**onsequences, are not directly determined by life events, i.e., **A**ctivating Events, but rather by the way we cognitively process and evaluate those events, i.e., **B**elief. Furthermore, this model states that it's not a simple matter of unchangeable process in which events lead to beliefs that result in consequences. Instead, it's the type of belief that's held that matters, and we have the power to change those held beliefs. Albert Ellis's model played a significant part in the form of therapy known as Rational-Emotive Behavior Therapy (REBT), which is like a precursor to the more commonly applied CBT. In REBT, beliefs are divided into "rational" and "irrational" beliefs. Using the ABC model, the aim is to help you accept the rational beliefs and dispute the irrational ones. The disputation process is what results in the model being referred to as the ABCDE model after it was upgraded to include these two steps. Applying it today might look something like this:

A: *Activating Event* (something happens to you or around you).

B: *Belief* (the event that causes you to believe either rational or irrational).

C: *Consequence* (the belief leads to a consequence, with rational beliefs leading to healthy consequences and irrational beliefs leading to unhealthy consequences).

D: *Disputation* (if one had held an irrational belief which has caused ill consequences, they must dispute that belief and turn it into a rational belief).

E: *New Effect* (the disputation has turned the irrational belief into a rational belief, and the person now has healthier consequences of their belief as a result).

Aaron T. Beck evolved and expanded on Ellis's works in the 1960s that contributed significantly to the modern CBT that we know. Beck noticed that many of his patients had internal dialogues that were almost a form of them talking to themselves. He also observed that his patients' thoughts often impacted their feelings and called these emotionally charged thoughts "automatic thoughts." Thus, he developed CBT as a newer form of therapy that looks at patterns and beliefs that can contribute to self-destructive behaviors. CBT can treat anxiety disorders, mood disorders, personality disorders, eating disorders, sleep disorders, psychotic disorders, and substance abuse.

WHAT'S THE PRINCIPLE BEHIND CBT?

Cognitive Behavioral therapy assumes that both the individual and the environment are of fundamental importance and that treatment outside of a holistic approach would be an injustice to the client. Three basic principles are underlying this approach:

1. CBT assumes that problems are based in part on unhelpful and unhealthy ways of thinking.
2. Those problems are rooted in part in learned patterns of unhealthy and unhelpful behavior.
3. An individual suffering from any psychological problem can learn better ways of handling them.

In so doing, the individual can relieve the symptoms and become more empowered and effective.

For these core reasons, Cognitive Behavioral therapy is about collaboration and participation. It emphasizes the present and requires an excellent client-therapist relationship. CBT is an ever-evolving formulation of the patient and their problems in cognitive terms, which aims to teach the client to be his or her own therapist. CBT sessions are time limited. Each session is carefully structured to aid in the successful execution of this form of therapy. It is goal-oriented and uses a variety of techniques to change thinking, mood, and behavior. From personal experience, I can see that CBT values and empowers an individual to take control of his or her problems and manage life in a healthy adaptive way. This is accomplished through psychoeducation.

3 WAYS IT CAN HELP YOU OVERCOME YOUR MENTAL ILLNESS

Three simple steps you can take to start implementing this form of therapy are:

The first step is to identify the negative thought.

Your therapist or someone you trust, and respect can help you uncover some of those unhelpful and unhealthy thoughts contributing to your current mental disorder. For example, using my story of how I started using CBT. It took me a while to see progress, but after a little effort, I realized that I had been carrying around thought patterns and feelings of unworthiness. I always felt like my fiancé was too good for me. Like I didn't deserve happiness or to be loved. And I quickly made the connection that I had always been carrying around these thoughts since childhood.

The same process will apply to you. While working together with your appointed therapist or guide, you can start talking about your feelings toward the current problem and name some of the dominant thoughts associated with those feelings.

The second step is to challenge that negative thought.

Once you've identified some unhealthy thoughts, the next step is to question the evidence for your ideas, analyze the beliefs behind those thoughts, and really challenge their validity. At this point, it helps to have someone objectively help you through this process. Then you'll discuss why you feel as you do and the corresponding behavior these thoughts have created. Finally, you'll need to test your negative thinking by separating your thoughts and feelings from reality. Often, we realize that our thoughts and feelings aren't based on facts when we go through this step.

The third step is to replace that negative thought with a realistic one.

If you change your negative thought to the extreme opposite, the new one won't stick in your mind for very long. Even personal development students have figured this out. When someone with a strong belief in poverty tries to shove the affirmation "I'm a billionaire" down their throat, it never yields any fruitful results. Saying "I am super-rich" when the belief of being poor is dominant might offer temporary relief at that moment, but it won't help you permanently shift your thought patterns, feelings, actions, or attitude toward wealth. When working with CBT to eliminate anxiety, depression, or whatever else, your therapy should help you formulate a new thought that is realistic enough for your mind and brain to accept as real for you. It's about creating new bridge thoughts that can ultimately get you to the desired final thought. The further away you feel from where you want to be in your thinking, the more you should create little bridges instead of trying to make a huge jump from where you are now to that new thought reality. So, for example, if you suffer from social anxiety, instead of saying, "I'm the coolest person in the world, and everybody loves being around me," a more constructive thought that will easily get anchored in your mind would be "just because it's awkward for me to be around people doesn't mean others see me that way." Build the new thought reality from that foundation, and it is likely to yield better long-term results.

This is the skill that Cognitive Behavioral Therapy gives us. It puts us in the driver's seat of our own lives and equips us with tools for navigating what would otherwise be overwhelming events. Since we know negative thoughts lead to negative feelings and actions, we reframe our thoughts into positive, constructive thoughts, leading to corresponding feelings and behaviors. Every mental health issue will be approached differently by your chosen therapist but let me share the foundational techniques that will always be included.

- You will identify specific problems in your current state and daily life.
- You'll become aware of the unproductive thought patterns and how they are impacting your life.
- The therapist will show you how to start reshaping your thinking in a way that changes how you feel about yourself and the problem or issues at hand.
- You'll learn new behaviors and begin putting them into daily practice.

Then you'll be advised to implement one of these techniques.

#1. COGNITIVE RESTRUCTURING OR REFRAMING

Do you have a tendency of assuming worst-case scenarios? Are you often expecting people to mistreat you, events to turn out wrong, or for you to mess up things that are important to you? Thinking this way affects almost everything you do. Fussing over minor details or getting too worked up and taking any small conflict personally will create a lot of disharmony in your life, and unfortunately, expecting the worst tends to become a self-fulfilling prophecy.

That's why it's crucial to identify your negative patterns and how you often react to situations. Once you become aware, you can reframe those thoughts, so they are more positive and productive.

For example, I was sitting at Starbucks with a friend soon after starting my journey to recovery. I had just spent countless hours studying cognitive-behavioral therapy. Because I had joined the college to major in psychology (before calling it quits), it wasn't too long before I started recognizing some of the dominant thought patterns that were ruining my life, like that one time sitting with John who was trying to comfort me after my breakup. The waitress brought his order and completely forgot about mine. It sounds ridiculous, but it's almost like I had been invisible the whole time, and everything she heard while we placed our orders evaporated into thin air. As you can imagine, when his coffee and eggs on toast came, I was infuriated. I told him, "See John, it's like women are out to get me. And every time I get a female waiter, something always goes wrong with my order." I was just about to flip the lid (he was used to it), but this time, a little voice stopped me from having a momentary outburst. And I just reflected for a moment on what I was saying and where that thought came from. Surely this girl wasn't out to get me. She had never seen me before. Why do I assume the worst all the time?

When I started having these "moments of awakening," things began shifting within me. That particular day, my friend was utterly shocked that I didn't have a temper tantrum with that waitress. Instead, I called her back and didn't even ask why she ignored my order. I simply asked her to pay close attention and retake my order.

#2. EXPOSURE THERAPY

This technique is used to confront fears and phobias. Your therapist should slowly expose you to the things that evoke fear or anxiety and offer guidance on coping with them in-the-moment. For example, if you suffer from social anxiety and the thought of being at a party full of people triggers a lot of fear, then you might want to incrementally expose yourself to how it would feel to experience that event. Break it up into chunks and experience different aspects such as arrival and greeting new strangers. Choose a technique that helps you get through that phase first, and when you can feel calm enough while replaying that scene, move into the next phase, e.g., striking a conversation with a few people who capture your interest. Keep working on it until you feel less vulnerable and more confident to cope with such an experience.

#3. RELAXATION AND STRESS REDUCTION TECHNIQUES

In cognitive behavioral therapy, you'll learn some progressive relaxation techniques such as:

- Muscle relaxation
- Deep breathing exercises
- Imagery

All these exercises lower stress and increase your sense of calm and control. They also help you deal with triggers in real-time, which gives you enough buffer time to prevent relapses.

#4. ACTIVITY SCHEDULING AND BEHAVIOR ACTIVATION

In this technique, you finally learn to deal with procrastination and other habits that are rooted in fear and anxiety. We help you do this by encouraging you to block time on your calendar for these anxiety causing activities. Once it's on your calendar, you are more likely to see it through, and it helps you develop the right habits needed to make you a high performer. Combining this with other techniques such as deep breathing can help you accomplish things that previously seemed impossible.

#5. JOURNALING AND THOUGHT RECORDING

I find writing an effective form of therapy, and now there's plenty of research to back this technique. You can start by listing the negative thoughts that hijack your brain and mood as they occur and then write out their opposite (or bridge thoughts) depending on how severe your case might be. A particular technique I regularly practice to-date is called clarity through contrast. Basically, I grab a piece of A4 paper and draw a horizontal line. On my left, I title it what I don't want, and on my right, I title it what I would love. I put my current thoughts or whatever ails me on the left side. All those ridiculous voices and feelings come out and on to that paper. Once I feel I have put out everything, I take three deep breathes and switch to the right side of the paper, filling in the thoughts and feelings I would love to feel and think. Most of the time, I don't even believe I can ever get to that other side, but just making this list of what I would love opens me up to a better feeling state and calms me down, which allows me to handle matters better.

Another exercise that works well, especially at the early stages of your healing, is noting down in a journal all the new thoughts and behavior you're putting into practice as you go through your therapy. Putting things down in writing helps you see your progress, which keeps you encouraged and focused on healthy thoughts and behavior.

#6. ROLE PLAY

Role-playing is an excellent technique if you struggle with expressing yourself and communicating with others. You can use this technique to practice social skills (e.g., if you want to ask a woman out but struggle with anxiety), improve your problem-solving abilities, gaining familiarity and confidence in certain situations, and so much more.

#7. GUIDED DISCOVERY

This is best done with your therapist or someone qualified who can listen to your problem and learn your viewpoint. Then he or she will ask questions designed to challenge your current beliefs. They will assist you in broadening your thinking by offering different perspectives. You might be asked to give evidence that supports your assumptions as well as evidence that does not. As you go through this process and open up your mind to see things from other perspectives, you become empowered to see and choose a new, more beneficial path.

#8. PROGRESSIVE MUSCLE RELAXATION

This technique is similar to the body scan (if you practice mindfulness, this should sound familiar). You do it by instructing your muscles to relax (one muscle group at a time) until your whole body gets into a state of relaxation. You can use audio guidance if doing it by yourself (you can even find videos on YouTube) or simply in a quiet relaxing ambiance with some candles or whatever stimulates relaxation for you. It can be beneficial when soothing a busy and unfocused mind or when you feel too nervous or anxious about something.

#9. BEHAVIORAL EXPERIMENTS

This technique allows you to become your own prophet in this sense. If you're fearful or anxious about something, ask yourself what you think is going to happen. What is it that is so bad that causes this reaction in you? Detail it out as much as possible. Then carry out your activity and take note of the actual outcome you experience. For example, if you're anxious about standing in front of an audience because you think you'll die or someone will throw rotten eggs at you because you suck, detail out that prediction, then go make your speech regardless if only to test whether your prediction was valid or not. Take note of how many of your predictions will come true. Over time you might discover that the predicted catastrophes that typically hold you back from things you want to do hardly ever happen. That alone is transformational therapy because it gives you the confidence and mental freedom needed to approach things with a healthy attitude and mindset. Ultimately, your anxiety to do things that are out of your comfort zone will dissipate. As Mark Twain once said, "I've had a lot of worries in my life, most of which never happened."

A PRACTICAL LOOK AT COGNITIVE BEHAVIORAL THERAPY

From a practical viewpoint, CBT is meant to be something short-term. It doesn't deny that your past is real but simply emphasizes the current present and what you can do to make things better for yourself. This talking therapy follows a simple model of thoughts (cognition) - feelings - actions (behavior), and by starting to make adjustments at that thought level, the end result is bound to change.

Take into consideration some of the techniques shared above. Techniques like deep breathing, journaling, and progressive muscle relation are simple yet profound practices that can help you alleviate some of the symptoms that afflict you and, in some cases, completely eradicate them from your life. The goal of cognitive behavior therapy is to teach you that you can control how you interpret and deal with situations and your environment even when the world around you is less than pleasant.

DOES IT WORK?

CBT has helped many people with certain types of emotional distress that don't require psychotropic medication. People suffering from depression, anxiety, anger issues, eating disorders, nightmares, phobias, addictions, panic attacks, and so much more. It's empirically supported and has been shown to effectively help patients overcome and heal a wide variety of life-crippling behaviors. The best part about this is that you can also combine it with medication if your case is chronic. You can get a therapist for more advanced CBT strategies or DIY if you feel capable. The techniques shared in this book are simple enough to start implementing now, making it one of the most affordable forms of therapy.

3

DIALECTAL BEHAVIORAL THERAPY (DBT): WHAT YOU NEED TO KNOW

P sychotherapy is one of the best forms of treatment methods for numerous mental health illnesses. So far, you've learned about the umbrella term CBT which is often referred to as talking therapy. For many of the common mental health problems, CBT works exceptionally well in helping the patient overcome their problems. However, not all patients can see the positive effects of this form of therapy (especially those suffering from borderline personality disorders and post-traumatic stress disorders). A doctor by the name of Marsha Linehan began to notice this back in the 80s. By the end of the 1980s, Linehan and colleagues decided to evolve CBT to create a more effective treatment for problematic, suicidal women. Linehan combed through the literature on efficacious psychological therapies for other disorders and assembled a package of evidence-based, cognitive behavioral interventions that directly targeted suicidal behavior. That was the birth of DBT.

Branching Out from CBT to DBT

DBT (dialectical behavior therapy) was initially focused on changing cognitions and behaviors of patients who felt criticized, misunderstood, and invalidated. Linehan weaved into the treatment interventions designed to convey the patient's acceptance and help them accept themselves entirely. That includes emotions, thoughts, the world, and others. This therapy was intended to treat borderline personality disorder, but it has since been adapted to treat other mental health conditions as well, including but not limited to eating disorders, substance use disorders, and post-traumatic stress disorders. As an extension of CBT, dialectical behavior therapy incorporates the philosophical process known as dialectics. The standard DBT treatment package consists of weekly individual therapy sessions, a weekly group skills training session, and a therapist consultation team meeting.

WHAT IS THE PRINCIPLE BEHIND DBT?

To understand DBT and its founding principles, you need to learn a thing or two about dialectics. So, what is dialectics anyway? It's a concept that states everything is composed of opposites and that change occurs when there is "dialogue" between opposing forces. The idea originates from ancient Greek philosophy, and according to Wikipedia, it is a discourse between two or more people holding different points of view about a subject but wishing to establish the truth through reasoned methods of argumentation. How does this become a helpful form of therapy for you?

Using this process, you are encouraged to resolve the apparent contradiction between self-acceptance and change to bring about positive changes in your life. There are three basic assumptions applied to this form of therapy. The first is that all things are interconnected. The second is that change is constant and inevitable, and the third is that opposites can be integrated to form a closer approximation of the truth. Another important aspect of DBT developed by Linehan is known as validation. Linehan and her team found that when validation was used along with the push for change, patients were more cooperative and less likely to quit or become distressed as they went through the changes. In practice, the therapist validates that a patient's actions "make sense" within the context of their personal experience without necessarily agreeing that the actions taken are the best possible approach to solving their desired problem.

HOW IS IT DIFFERENT FROM COGNITIVE BEHAVIORAL THERAPY?

From the name, we can tell that CBT and DBT share some similarities. DBT evolved from cognitive behavior therapy, but its approach is distinct enough to merit being considered a unique model. Both CBT and DBT are supported by extensive evidence-based research. So, we can feel confident about their effectiveness. They can each be used to treat a wide variety of mental health problems, but in some instances, one is more suitable than the other.

That's why it's good to understand some of the differences between these two therapy models.

The main difference is the kind of change they create for the patient, according to scientific research. CBT can help a patient recognize and change problematic patterns of thinking and behaving. On the other hand, DBT is best when a patient needs help regulating intense emotions and improving interpersonal relationships. Through validation, acceptance, and behavior change, DBT can help a patient create those required shifts. In DBT, there's no heavy reliance on changing thoughts. There's an implicit process that happens so that as the client is mindful, more accepting of themselves, and as they learn to validate themselves and ask for validation, they start to change any resistance they may have. Ultimately, they become kinder to themselves, get more grounded in reality, and accept reality without catastrophizing everything in their world. But they don't go through the process of actively challenging their thoughts like in CBT.

One thing to note here is that research doesn't say there's a one-size-fits-all. Although the doctor who evolved DBT was focused on suicidal patients, this model has worked for other mental health disorders. So, we must remain open-minded enough and avoid making the mistake of assuming that one is better than the other. Mental health disorders affect cognition and behavior differently, so neither CBT nor DBT is the best option in all cases. You need to figure out which is the best option for your particular case.

Research shares so far that depression, anxiety, OCD, phobias, and PTSD are typically better approached through cognitive behavior therapy. For borderline personality disorders, chronic suicidal ideation, and self-harm behaviors, dialectical behavior therapy is usually more effective.

THE STRATEGIES OF DBT THAT HELPS TRANSFORMS NEGATIVE BEHAVIORS INTO POSITIVE ONES

Using DBT, you get to learn and develop four core skills to cope with emotional distress in healthy, positive, and productive ways. These skills are sometimes referred to as modules and are considered by Linehan as the "active ingredients" of successfully utilizing DBT. They are mindfulness, emotion regulation, distress tolerance, and interpersonal effectiveness. Let's examine each skill.

Mindfulness

What is mindfulness? Mindfulness means maintaining a moment-by-moment awareness of our thoughts, feelings, bodily sensations, and the surrounding environment through a gentle nurturing lens (definition from greatergood.berkley.edu).

The roots of mindfulness can be traced back to Buddhism, more specifically, Buddhist meditation. Today, most people only know of the more secular practice of mindfulness, which entered mainstream America in the late 1970s. Many studies have been documented since the 70s to show the physical and mental health benefits of adopting mindfulness practices.

We also need to consider the aspect of mindfulness that is about acceptance. What do I mean? In mindfulness, we are encouraged to pay attention to our thoughts and emotions without judging them, without believing even for an instant that it's wrong or right for you to feel how you're feeling in any given moment. For example, as you're going through this chapter, notice how you feel in this moment as you read my words. If you're feeling soothed and encouraged by what you're learning, that's great, and you can embrace those thoughts and emotions. If you're struggling to agree with or even see value in what you've learned so far, perhaps you're starting to get frustrated and impatient with me; it's also okay. As you begin to notice what's happening within you and how you respond to the activities and interactions you have throughout the day, you begin to practice mindfulness. That helps you tune into what you're sensing in the present moment. It also keeps you grounded in the NOW instead of being pulled back into the past or dashing into an imaginary future.

In the context of using this practice in DBT, acceptance plays a significant role. The more you notice and accept your thoughts and feelings through that nurturing less, the easier it becomes to progress with your healing. So DBT breaks down mindfulness into the "what "skills and the "how" skills.

The "what" skills are about noticing what you're focused on. That includes the present moment, your awareness in the present, your thoughts, emotions, and sensations, and separating emotions and sensations from thoughts.

The "how" skills teach you how to be more mindful. You do this by learning to balance rational thoughts with emotions, using radical acceptance to learn to tolerate aspects of yourself as long as they aren't harmful to you or others. It's also about taking effective action and overcoming things that make this practice of mindfulness challenging to execute. Things like restlessness, fatigue, sleepiness, and doubt all inhibit your ability to practice mindfulness, so they need to be managed effectively. And lastly, you need to use these mindfulness skills that you learn regularly. It cannot be a one-time thing or only when it's convenient for you.

Emotion regulation

While mindfulness practices are extremely powerful, we know that sometimes it's just not enough, especially when you feel like your emotions have a life of their own and you've been kidnaped by them. For those of us who've battled with mental health issues long enough, we know how crippling emotions can be. There are moments in my life when I felt like there was no escape from the hellish experience of my feelings. If you've had those moments too, I feel for you. And I want you to know that as difficult as it might be, it's possible to manage them. That's what this skill of emotion regulation teaches you.

Emotion regulation teaches you how to deal with immediate emotional reactions before they lead to a chain of distressing secondary reactions. You will learn how to recognize emotions, reduce your sense of vulnerability and insecurity. You'll overcome barriers to emotions that have positive effects and increase those emotions with positive effects. That will help you become more aware and mindful of your feelings without judging them and expose you to your emotional self. The more exposed you are to your emotions, the easier it will be to avoid giving into emotional urges because you'll be able to recognize them and that in turn will solve problems in helpful ways. In case you're wondering what, a primary emotion might be, let me share an example.

Suppose you're about to get extremely angry for whatever reason. Still, because you've practiced emotion regulation, you get a buffer time that enables you to recognize that you're about to explode with anger. In that case, you can practice these core skills that you're learning in this chapter (depending on what feels most effective in that situation) and calm yourself down. That will enable you to avoid entering into a downward emotional spiral, which might lead to guilt, unworthiness, shame, and even depression depending on how bad things would get if you didn't regulate yourself while in that particular situation.

Distress tolerance

While regulating your emotions is wonderful and prevents you from falling into bouts of depression and other terrible states, some environments are just rigged with triggers that seem out to get you. In a moment of crisis, even with all your mindfulness and commitment to regulating your emotions, how can you best avoid that pit of despair if nothing else seems to be working? The short answer is by activating distress tolerance.

You learn distress tolerance skills so you can get through moments of crisis and rough storms without the need for potentially destructive coping techniques. That's why I love this module. Before developing distress tolerance skills, I would be fine until something unexpected throws me off my game. Then I would try to use coping mechanisms like self-isolating, avoidance, and alcohol to deal with situations that I just couldn't control. Suffice to say, the results weren't what I wanted. Many have had a similar experience, so I encourage you to learn these skills instead of seeking temporary coping mechanisms that might cause more harm. By learning distress tolerance skills, you learn to distract yourself until you become calm enough to deal with the destructive emotion or situation. You learn to self-soothe by relaxing and using your senses so you can reconnect to that feeling of peace. These skills are priceless because they always help you find ways to improve the moment even if you're in great pain or difficulty. If you must use some coping technique, you won't blindly pick the easiest one. Instead, you'll be mindful and thoughtful about the pros and cons of your choices. Using this technique, I recognized that drinking as my coping mechanism was only making my life worse. I could catch myself in that moment when the idea of getting a drink popped into my head and would reach for a better feeling solution instead, like calling my support group or going for a jog to get the steam out. That's when things really started shifting permanently. I got to a point where I stopped being fearful and anxious about relapsing into depression.

Interpersonal effectiveness

Relationships are always going to be integral to our growth and happiness as human beings. No matter how heartbroken you feel or how betrayed you felt, learning how to better connect and relate to others is still important, and it should be part of your recovery process. Interpersonal effectiveness skills help you view relationships from a new lens. It gives you clarity and perspective. These skills combine listening skills, social skills, and assertiveness training so you can learn how to change situations and still remain in integrity with your values. These skills include learning how to work through conflict and challenges in relationships and learning how to ask for what you want. It's also about building great respect and love for yourself.

WHAT TO EXPECT IN DBT

DBT uses specific techniques to achieve the treatment goals that will help you get better. It usually includes a combination of individual sessions and group support.

Individual therapy typically comes first.

In these weekly sessions, the emphasis is on recognizing and self-monitoring your thoughts, emotions, and behavior by using a form of diary card that will be processed by your therapist. This card helps you and your therapist keep track of the treatment goals. The goal here is to bring about greater awareness of your triggers, thoughts, emotions, behaviors, and actions so that you can elicit change strategies. It is usually a 50-minute session. During individual therapy, you will also work on emotion regulation, traumatic experiences, and any other issues that arise.

You can expect your therapist to be active in teaching and reinforcing adaptive behaviors between and during sessions. The emphasis should be on teaching you skills that will empower you to manage your emotional trauma. It's about working together with your therapist so you can learn to improve many of the necessary social and emotional skills needed to help you feel more in control. You might also receive some homework assignments during these individual sessions.

Weekly group therapy sessions are the second component of DBT.

These sessions are typically two and a half hours long. These structured gatherings led by your therapist will help you learn, incorporate and practice one of the four different modules discussed earlier. After all, unless you can practice these core skills, it won't stick or produce lasting effects in your life. In the group, you also get the chance to discuss homework assignments, practical applications of the new skills you're learning, etc.

During your DBT treatment, you will have access to your therapist by phone or virtually if you need help dealing with any crises that show up.

As you commit and go through this process, you will start to experience the benefits of learning skills that improve your tolerance to distress and emotional regulation. If you were dealing with any self-harming behaviors or suicidal thoughts, these would be addressed first. And as you combine the individual psychotherapy sessions with the supportive group experience, you'll start practicing more of the interpersonal skills you'll learn from your therapist, decreasing maladaptive behaviors and thoughts affecting your quality of life and relationships. In short, you'll finally start getting your life and peace of mind back in order. As your self-confidence, self-respect, and self-belief increases, you'll learn how to set reasonable goals to improve your lifestyle.

DBT, ARE YOU THE ONE FOR ME?

How will you determine whether DBT is right for you, and can you be sure it will work?

Well, some signs and symptoms would make DBT seem like the best therapy form. Do you suffer from any of the following?

- Self-destructive behaviors such as alcohol or drug abuse, binge eating and purging, sexual promiscuity, and other impulsive behaviors like gambling, gaming, or spending sprees.
- Repeated suicide threats or attempts.
- Chronic problems with depression, anxiety, and anger.
- Self-harm behavior such as cutting, burning, and picking yourself.
- Hypersensitivity to criticism, rejection and disapproval, fear of abandonment, and a pattern of unstable interpersonal relationships.
- Intense and volatile emotional reactivity and difficulty returning to a stable mood.
- Poor and unstable self-image with a strong sense of emptiness.
- Feelings of paranoia and victimization.
- Detached thinking that ranges from difficulty maintaining attention to episodes of complete disassociation.

If you are struggling with one or more of these issues, DBT can be the right one for you. There are many success stories where using Dialectical Behavior Therapy has been the ultimate treatment that helped avert suicide. In fact, one of the best stories I can share with you is that of Dr. Linehan, the woman behind this form of treatment. In 1961, at the age of 16, she was admitted to a psychiatric hospital. She was there for over two years, and almost all the therapists that came to treat her proved ineffective. She says some of them were so bad, she often felt like they were driving her deeper into what can only be described as hell. She considers it a miracle that she was able to heal and make a successful life for herself, getting a Ph.D. in social and experimental psychology. Out of that, she was able to evolve existing therapy treatments into the kind of treatments she once needed. "I developed dialectical behavior therapy (DBT) for people who suffer unimaginable emotional pain and resort to desperate behavior amid suffering. The people I most wanted to help were those at very high risk for suicide, and DBT turned out to be extremely successful in helping suicidal people stay alive." Dr. Marsha M. Linehan.

HOW DAVID HEALED HIMSELF FROM PAIN MEDS ADDICTION

David is a friend I met years ago in my support group, and we've continued our friendship long past our old lives when everything seemed unbearable. Our connection might be because we both struggled to recognize that something was terribly wrong and that we needed help until it was almost too late. Although he didn't get heartbroken by a woman, he still underwent his own version of a living hell. According to David, he couldn't remember a time in his life when he actually felt like life was worth living. What's worse is that he was referred from one specialist to another, all of whom seemed to only bury him deeper in his depressive state. He was also prescribed various psychotropic medications, which had severe side effects.

To make matters worse, his depression was getting more chronic. So, to make things a little more bearable, David started soothing himself with some strong pain meds his mother was taking. It wasn't too long before that became his new coping mechanism, and over time, he became a serious addict. That's when his mother decided to try a different form of therapy. By the time his mom was able to get him into DBT, he was suffering from Major Depression and Substance Abuse Disorder. He was also starting to develop certain self-destructive behavior that his mother worried would lead to suicidal threats. The treatment that David reluctantly joined is what connected us, and the rest, as they say, is forgotten history.

After going through DBT, David was able to develop new skills, heal, and transform his life. There was a time in his life when he couldn't get out of bed for days on end because of the despair and emotional suffering he was undergoing. Today, David is a successful entrepreneur running his own business, and he volunteers at his local community to pay it forward. He's also started a YouTube channel to encourage men suffering from mental health issues to seek out support and treatment because he feels it was more of a struggle for him due to his reluctance to speak up. He tried to ignore the warning signs and didn't want to appear "weak," but now he recognizes the importance of getting help when one feels mentally unwell.

4

A CLEAR OVERVIEW OF ACCEPTANCE & COMMITMENT BASED THERAPY (ACT)

Researchers continue to work tirelessly to develop and discover the most effective ways to help people overcome mental challenges. Long-term recovery and relapse prevention still remain a huge obstacle that many strive to resolve because there's nothing worse than putting someone through a program successfully only to have them relapse shortly after. There's also the issue that not all standard therapy programs effectively help people overcome psychological pain. That's where ACT comes in.

GETTING INTO THE ACT

Acceptance and commitment therapy (ACT) is a new form of treatment developed more recently (in the 90s) with the hope of increasing long-term success in the treatment of mental health conditions. It's based on relational frame theory (a school of research focusing on human language and cognition). This therapy is more action-oriented and also has its roots in traditional behavior therapy and cognitive behavioral therapy. ACT teaches you how to stop denying, avoiding, suppressing, and struggling with your inner emotions and instead start accepting that you are not flawed or broken. These negative, often shunned feelings are appropriate responses to specific situations, but that should never inhibit you from moving forward in your life. In other words, ACT enables you to accept your issues and hardships and to commit to making necessary changes in behavior regardless of what's going on around you. Treatment using ACT has been successful when applied to chronic pain, eating disorders, depression, psychosis, anxiety, and substance abuse.

WHAT'S THE BASIS OF IT?

At a fundamental level, ACT is concerned with helping you realize the fullness and vitality of your life. It's about bringing new meaning to your life and helping you discover your values so that you can lead a value-based lifestyle that ultimately brings a greater sense of worth. It also emphasizes the importance of embracing that a full life includes a broad spectrum of emotions and human experiences, including pain. Acceptance of things as they come without evaluating or attempting to change them is one of the core skills you learn, making it easier to shift and manage painful or difficult situations.

THE 3 KEY FUNDAMENTALS TO ACT

#1. Acceptance

Accepting one's emotional experience can be described as the process of learning to experience the range of human emotions with a kind, open, and accepting perspective. Whether it's a situation you cannot control, a personality trait that's hard to change, or an emotion that overwhelms you, ACT invites you to accept reality and work with life as it is instead of trying to fight or control things beyond your control.

#2. Choosing the Desired Direction

You get to choose the best direction that you can take based on your values during this process. To do this, you'll be guided through a values clarification process that will help you define what's most important to you. Your mental health professional will share exercises to help you identify your core values then you can align them with your actions to respond to pain and difficulty from that frame of mind. It's also a great way to clarify how you wish to live life and what's meaningful to you.

#3. Taking Action

The last key is taking action and committing to the changes you've made in behavior and moving in the direction of your identified values. These three keys are interconnected and cannot be treated as separate when going through the therapy. What underlies all of them is that each one must be approached through the lens of mindfulness. So, let's touch on the role of mindfulness in the ACT.

During ACT, mindfulness will allow you to connect with the observing self (the part of you that is aware but separate from your thinking self). ACT uses mindfulness so that you can detach from thoughts and experiences that make it hard for you to see clearly or even remain present. Challenges related to painful feelings and past situations that no longer serve you are first reduced and eventually accepted through mindfulness practices. So, for example, if like me you struggle with thoughts of unworthiness and you've had a painful experience like what I had where someone I loved betrayed me in the worst possible way, which had me believing that I really am unworthy and unlovable, then instead of replaying that thought "I'm unworthy and unlovable," you might be asked to instead say "I'm having the thought that I'm unworthy and unlovable." This effectively separates you (the individual) from the cognition, thereby stripping it of its negative charge.

The Single Key Difference That Separates ACT From Other Therapies

Thus far, you've learned about three types of psychotherapy treatments that all work in helping people overcome mental health problems. CBT (cognitive behavior therapy) is the foundation of both DBT and ACT. However, they are all distinct in their own ways, which is why they are considered separate. When it comes to understanding the core difference between ACT, DBT, and CBT, the key difference is that ACT emphasizes facing painful emotions and experiences head-on and taking positive action forward. It focuses on embracing, accepting, and transforming the feelings through a committed effort to shift one's behavior and perceptions about the problem. It's also more focused on accepting oneself and seeing the good instead of believing that we are flawed in some way for developing mental issues. Through acceptance and positive action based on one's values, tackling life's difficulties can become manageable, and life can once again become fulfilling and enjoyable.

PRACTICAL STEPS IN APPLYING ACT

Whether you have a therapist guiding you through the acceptance and commitment therapy or not, you can still apply some of the exercises to your current situation. Many of the mindfulness exercises are practical and easy for anyone to implement. For example, if you want to practice Acceptance, a simple exercise to do is "Opening up."

If unpleasant feelings are showing up for you right now, see if you can take a few deep breaths and just allow them to be there with you. Instead of suppressing or sweeping them under your mental carpet, embrace these feelings. Explore what there is to experience and notice your body's sensations and the thoughts, images, and emotions running through your mind. Can you stay present with these difficult feelings and keep in touch with them without judging yourself? Do these feelings remain the same or do they start changing as you allow them non-judgmental space to just be with you? Stay with them a while longer and notice whether there's any fluctuation. Are they getting heavier, lighter, or still the same? Notice how you're talking to yourself as these feelings are experienced. What interpretations are you making about your experience, and is it really based on reality?

Now, see if you can counter some of the negative self-talk with more realistic ones and then re-evaluate that experience from your newly found perspective.

Other practical strategies that can be useful in your daily life include:

- Acknowledge the difficulty in your life without escaping from it or avoiding it.
- Give yourself permission to not be good at everything.
- Allow feelings or thoughts to happen without the impulse to act on them.
- Realize that you can be in control of how you react, think, and feel in any given situation.

NO NEED TO ACT, ACT WORKS!

ACT was developed in the late 1980s by Steven C. Hayes, and most academics like to consider it the "third wave" of behavioral psychotherapy approaches. Think of it as an updated version of CBT, yet with more mindfulness and present-moment processing. There's a lot of scientific evidence that mindfulness practices and cognitive behavior therapy positively affect treating mental health problems such as anxiety, depression, substance abuse, and trauma. With ACT, metaphors, paradoxes, and experiential exercises are frequently used. Many interventions are playful, creative, and innovative ranging from short ten-minute interventions to some that extend over many sessions. An ACT-informed therapist usually takes an active role in guiding the client by exploring their values and building skills associated with mindfulness. For example, I have a friend (let's call her Ann to protect her privacy) who recently completed an ACT program. She shared that during her sessions, she felt guided into developing more compassion and mindfulness. She changed her relationship to her thoughts and started to feel the shift after just a few sessions.

Ann initially sought help because she felt so dissatisfied with her life, but at the same time, she was beating herself up and feeling guilty for having these thoughts and emotions. According to the status quo standards, Ann has a great life, and her husband is highly successful. Her days are occupied by their three young children, and although she has the lifestyle most housewives only dream about, she still felt dissatisfied. Not wanting to share this with her family for fear of being thought of as selfish and ungrateful (especially her traditional southern mom), she decided to seek professional guidance. Through ACT, she was guided to explore her values through a values exercise, which enabled her to examine whether or not she was living within her value system and in what ways she might desire to change things. Through this exercise, she has finally accepted her dissatisfaction and figured out ways to shift her current routine so she can live more aligned with her own values.

Besides personal stories that I have encountered of the success and viability of ACT as a form of treatment, there's enough scientific evidence backing up the validity of this form of psychotherapeutic approach. One such evidence is the result shared in the European Journal of Psychiatry Volume 32, Issue 4. Sixty-seven inpatients of a German psychiatric department were assigned to either ACT or

CBT+ condition assessed with respect to symptom measure and ACT-specific outcomes. The results showed that both groups improved on measures of symptom severity as well as ACT-specific components. There were no significant between-group differences. (Source. Effectiveness of Acceptance and Commitment Therapy compared to CBT+: Preliminary results).

II

USING THE PRACTICAL METHODS IN CBT, DBT, AND ACT IN YOUR DAILY LIFE

5

YOUR IRRATIONAL THOUGHT PATTERNS AND RISING ABOVE THEM

Have you ever been in a situation where someone is talking to you, and you get an image in your head of punching them till they bleed? Or perhaps you're just minding your own business while having your morning coffee at Starbucks, and all of a sudden, you start thinking about what everyone around you would look like naked. As Rihanna would say, these are "wild thoughts," and it is quite normal to have an irrational thought pop up every once in a while. In fact, I think every human experiences some fleeting irrational thoughts in some way, shape, or form at some point in their life. It usually comes and disappears quickly.

The problem is when the thoughts are persistent, and they hijack your attention and emotions. If you keep catching yourself drowning in an inner dialogue or mental movie scene that feels out of control and is rooted in harmful or destructive ideas, then you've landed in the realm of distorted thinking.

Irrational thoughts can be self-directed or directed toward others. Regardless of whom they are focused on, these thoughts are usually sad, disturbing, negative, and at times destructive in nature. Letting them hang out in your mind is often the reason stress, anxiety, and depression accelerate, so it's best to become aware and to devise healthy ways of dealing with them as they show up.

IRRATIONAL THOUGHTS AND WHY WE HAVE THEM

The first thing to realize is that we have thousands of thoughts each day. Out of these, the majority are recycled thoughts. The implications of knowing this is that many irrational thoughts aren't originally yours. They are simply recycled, and you just happened to pick them up and call them your own.

There are many different types of irrational thoughts (expertly known as cognitive distortions in CBT), but at a fundamental level, they all seem to be unrealistic and definitely harmful to your mental well-being. These thoughts might include unjustified worry of financial hardships, fear that no one likes you, that you're unlovable and will always be alone, they might be thoughts of harming yourself or others. It could be a persistent thought of others falling ill or dying. As they come up, they seem to hijack your senses and attention, which sends you spinning off on a downward spiral with no hope.

From what research shows, we tend to have these irrational thoughts when under emotional distress. Strong, intense fear or anger is usually the culprit behind cognitive distortions. Pessimists and people who hate change typically struggle with irrational thinking.

HAVE YOU HEARD OF COGNITIVE REFRAMING (CR)?

Cognitive reframing, sometimes called cognitive restructuring, is a therapeutic process that helps the client discover, challenge, and modify, or replace their cognitive distortions. This is a staple tool in cognitive behavior therapy used frequently by therapists because many of our problems stem not so much from the crisis or unexpected situation but from the faulty way of thinking that we have about ourselves and the world around us. Through cognitive reframing, you will finally learn to reduce stress in your life as you develop healthy ways of restructuring your thought patterns and cultivate more positive and functional habits. Of course, this is easier said than done. Anyone who has suffered mental health issues and irrational thinking can attest to the fact that changing one's thoughts can feel almost impossible. But can you recall a time when doing something felt overwhelmingly hard? Perhaps it was learning to drive or getting through pre-calculus or any other skill you've picked up throughout your life. At first, it does feel strenuous, but just like any other skill, the more you practice, the easier it becomes. Challenging your own negative thoughts and beliefs and even catching yourself when your emotions are getting hijacked by distorted thoughts becomes manageable. We want to use this tool to develop the curiosity of what's real whenever thoughts and images hijack our mental space. When looking for a solution to overcome mental health issues, a good rule of thumb is to know that automatically trusting all your thoughts might not be the best idea. So, with a tool like cognitive reframing, you have the ability to test those thoughts for accuracy so you can better respond to the situation.

GETTING TO KNOW THE IRRATIONAL PATTERNS

The ways in which irrational thoughts play tricks on us are so numerous it's almost laughable. These tricks are called cognitive distortions in psychology, and you must start recognizing some of the patterns and beliefs that are playing out in your life. These beliefs and thought patterns may seem real, but they won't stand the test of light when analyzed accurately. They make your recovery hard and relapse after treatment a real possibility because as long as they are dominating your mental space, they are silently causing damage to your well-being. Unfortunately, many of these are ingrained during our formative years and feel like part of who we are. But that's false information, and it's crucial to start catching yourself when these culprits are threatening your well-being. Here are a few you should become aware of:

Catastrophizing

Are you one of those people who is always expecting the worst to happen? For example, you're sitting at a restaurant waiting for your date to arrive. You've been looking forward to this evening for such a long time. And then you realize your date is running late. Twenty minutes into it and still waiting. There's no answer when you call. It just rings and rings, so you automatically assume the worst and leave the restaurant. Of course, that reinforces the thoughts in your head that you're not loveable; she was playing you all along and never even liked you, etc. You start thinking about how worthless you are and how much women suck!

Irrational thoughts of the girl out with another man and laughing about you and what an idiot you are for thinking that she would ever go on a date with you fill up your mind. No matter what soothing words or solutions you try to apply, there's no remedy. This is an example of catastrophizing. It's about seeing only the worst possible outcome in everything.

Magical thinking

This distorted thought pattern is most common in children and adults with obsessive-compulsive disorder. Magical thinkers believe that they can avoid harm to themselves or others by doing some sort of ritual. Sometimes these rituals pose a threat to the person, but they don't see it that way. You might also find some bipolar cases with magical thinking tendencies.

Paranoia

In its extreme forms, paranoia sips into the realm of delusion. Many bipolar people experience less severe forms of paranoia because of personalizing events, catastrophizing, or making leaps in logic. If you've had thoughts that make you feel like everyone at the cocktail party you went to was watching and judging you, then you've probably experienced this. The truth is, most of the time, people are too consumed in their own issues to judge us in the way we tend to think.

Minimization

I see this one as the opposite of catastrophizing because it's about devaluing ourselves or the things that happen to us. It can apply to oneself or in relating to others. For example, if a person fails to meet your high expectations in one way, like telling a lie on a single occasion, you will now write that person off forever, refusing to see any good characteristics that may exist. Another way this expresses itself is through refusal to see the good or bad qualities in yourself and others.

Personalization

As the name implies, this is about taking everything personally and assigning blame to yourself without any logical reason. It can be as simple as blaming yourself when that date didn't show up for dinner or as severe as believing that you are the cause of every bad thing that happens around you. You might even call yourself a bad luck charm.

Control fallacy

A control fallacy manifests as one of two beliefs. Either you believe you have no control over your life and are nothing more than a helpless victim with a pre-determined fate or that you are completely in control of yourself and your surroundings and thus responsible for the feelings of those around you. Both beliefs are equally inaccurate.

All or nothing thinking

In this type of perception, you only see things through the binary black and white lens. Life has no shades of gray, and you are either an abject failure or a complete success. There's absolutely no in-between for you, which can often lead to despair and major misperceptions about your life and goal achievement.

These are but a few of the various irrational thought patterns that might be dragging you down and inhibiting your full recovery despite all your efforts to get better. While these distortions are common and potentially extremely damaging, you don't have to live with them any longer. There are various techniques you can use to identify, challenge and erase or at least minimize them. The first step is, of course, increasing your awareness of your thoughts.

THE NEXT STEPS TOWARDS A POSITIVE OUTLOOK

To overcome cognitive distortions, the first thing you need is a healthy way of identifying and understanding your irrational thinking. A useful tool you can use is an automatic thought record.

Create a six-column worksheet for yourself either in your private journal or Google document. The six columns should be date/time, situation, automatic thoughts (ATs), emotions, your response, a more adaptive response. If you need something more flexible, you could also use your smartphone's notebook so you can record throughout the day.

First, you need to record the date and time of the thought you're recording. You can do an hourly check-in or something more frequent.

In the second column, write down the situation you just experienced. Ask yourself:

1. What led to this event?
2. What caused the unpleasant feelings I am experiencing?
3. Who was involved?

In column number three, you will write down the negative automatic thoughts that came up, including any images that accompanied the thought. You will consider the thoughts and images that went through your mind, write them down, and determine how much you believed these thoughts. Examples can include *I'm such an idiot, no one loves me, I can't cope with this, I'll never get better, I'll never find another relationship, the world is always working against me, I'll never find a job, I suck at everything I do, etc.*

The purpose of this exercise is to find the "hot" thought or the one with the most significant electrical charge that messes with your entire system. That thought is the one that needs to be worked on first.

After identifying the thought, take notice of the emotions and sensations that accompanied said thoughts. This should go into the fourth column. What emotions did you experience at the time, and how intense were they on a scale of 1(barely felt it) to 10 (completely overwhelming). Examples can include anxious, guilty, ashamed, depressed, afraid, helpless, angry, happy, etc. You might wonder why I've included "happy" in this example list, but it's because I want to show you that thought recording is about getting in touch with and observing all your emotions at that designated time. So, you need to monitor both the negative and positive thought patterns.

In column five, I invite you to come up with an adaptive response to those thoughts. This requires courage and some effort on your part as you identify the distortions that are cropping up and challenging them.

Answer the following questions here.

1. Which cognitive distortions were you employing?
2. What is the evidence that the "hot" thought(s) is true, and what evidence is there that it's not true?
3. You've thought about the worst that could happen, but what's the best that could happen? What's the most realistic scenario?
4. How likely are the best-case and most realistic scenarios?

Since your current thinking is biased to that negative thought, this step should be easy. I want you to do your best to stick with verifiable evidence such as data, percentages, facts, and real proof. Avoid opinions and interpretations. A simple example to help illustrate this could be:

Suppose it was you who was waiting at a restaurant for over 15 minutes for your date to show up, and she didn't pick up the phone when you rang to find out where she was. When working out this process after that event, credible evidence that would be considered acceptable are:

- I was sitting at the restaurant for over twenty minutes, and she didn't show.
- I rang her cellphone once, and there was no answer or response.
- I was disappointed she went cold and didn't even call to explain that she wanted to cancel.

However, the following statements are not proof of evidence:

- She hates me and wanted to humiliate me in public.
- She had no intention of going out with me since the beginning.
- She's ruined my chances of finding love.
- I will never find a woman who genuinely loves me.

In the last column, I want you to consider the outcome of this event. Think about how much you believe the automatic thought, now that you've come up with an adaptive response, and just rate your belief out of ten. If you feel the grip of that initial thought lessening, then the process is working. Write out what emotions you're feeling now and at what intensity you're experiencing them.

ADDITIONAL TECHNIQUES THAT YOU CAN PRACTICE

To avoid being hijacked by irrational thoughts, I encourage you to practice at least one of these methods.

#1. Personalization

This cognitive distortion is about seeing yourself as the cause of all negativity around you, including others' misfortunes or everyday mishaps. It can take many forms. For example, you made dinner reservations for yourself and friends, but your name isn't on the list when you show up. Or, in my case, I'm sitting with a friend at a coffee shop, and the waitress gets his order right but not mine. Another scenario could be going on a beach vacation with your family only to have it rain the entire time. Yet on the last day, as you head to the airport, it's sunny clear blue skies. If you recognize that you usually take things personally when something happens, or someone says or doesn't say what you want, here's how to challenge those irrational thoughts as they come up.

How to handle it:

Notice how you're quick to take responsibility for something that's out of your control. Ask yourself, "could I factually control or contribute to this problem and how?" Then consider all the other factors that may have contributed to the problem. For example, in the case of the dinner reservations with friends, perhaps you could have double-checked they got the right date and time and that they didn't misunderstand your information. Could there be a glitch in the software system, or did the person who took the reservation forget to confirm it? When it comes to rain during a vacation, did you really cause rain just because you wanted good weather so badly? What facts can prove that you are the rainmaker and that it's your fault? Get curious, not judgmental, and have an open dialogue with yourself about these thoughts.

#2. All-or-nothing thinking

The do or die mentality will not serve you in most cases and only causes distress as you move through life because, in reality, life isn't binary. People aren't binary, and it will serve you well to question those thoughts whenever they come up.

How to handle this:

Notice the times when they come up and question if there's no other possibility other than that extreme thought trying to hijack your mind. For example, if you're thinking you didn't get hired at that interview because you're the worst, perhaps you can introduce the thought, "Is that really the only reason they didn't hire me? Couldn't there be another possibility? What if there's another reason?"

I like to play "what if there's another reason? What could it be?" when this type of irrational thought comes up. I find it breaks the doom and gloom pattern, enabling me to process the situation from a new perspective.

#3. Decatastrophizing

This tool is a great one to talk yourself out of the habit of catastrophizing situations or seeing the worst in people.

How to apply this:

When that worst-case scenario shows up, take a deep breath, find your journal or a piece of paper, and then write down your worry. Identify the core of the issue. What are you worried about? Do your best to identify the actual problem causing you to think this way. Now, picture for a moment how horrible it would be if what you're thinking actually came to pass. That worst-case scenario that you're so afraid of. Does it feel better thinking about this? When have you ever even experienced that same event or something similar in the past? How often has it happened? If it doesn't feel good playing out this worst-case scenario, and since you're not 100% certain that it will actually happen even if it did in the past, why not consider a different outcome. What if something good happened instead, and you got the opposite of that unpleasant outcome. What would that look like? Invest some time painting that scenario in your mind with as much color (if not more) as you did when thinking about the negative outcome. Sit with it until your emotions catch up. Consider the details of this new scenario and write them down. How does it feel to sit with this outcome?

Next, think about your chances of surviving in one piece. How likely are you to be okay if your fear comes true? How are likely is it that you'll be okay in one month or one year?

Finally, come back to the present moment and think about how you're feeling now. Are you still just as worried, or did the exercise help you think a little more realistically? Write down how you're feeling about it.

#4. Facts or Opinions exercise

I find this one to be incredibly therapeutic and use it all day, every day. One of the first lessons you learn in CBT is that facts are not opinions. These might seem obvious for people without mental health issues, but you and I know how hard it can be to remember and apply it in day-to-day interactions. I encourage you to start exercising with facts and opinions as you go through your day. To help you practice this immediately, here are some of my statements that I began playing with years ago. First, I made a long list of the common thoughts that dominated my day at the time, then I would sit for five minutes three times a day and go through each of them, labeling them either fact or opinion to help me remember what's real and what's not.

You can borrow my list or make one of your own.

- I am a failure.
- No one likes me.
- I am uglier than [name him/her].
- I suck at everything.
- I'll never find love again.
- I will never get better.
- She didn't care about hurting me.
- No one understands me in this world.
- I'm a terrible person.

- Bad things always happen to me.
- I am so unlucky.
- This will be an absolute disaster.
- Nobody could ever love someone like me.
- I ruined everything.
- I'm too fat.
- I'm selfish and uncaring.
- I can't do anything right.
- I'm too old.
- It's too late for me now.
- I ruined the evening.
- I failed the exam.
- A friend in need said "no" to me.
- I made a mistake and caused us to miss the movie.

Once you have your list of the most common thoughts, go through each, identifying which is fact and which is an opinion. Here's what mine looked like.

- I am a failure. *False*
- No one likes me. *False*
- I am uglier than [name him/her]. *False*
- I suck at everything. *False*
- I'll never find love again. *False*
- I will never get better. *False*
- She didn't care about hurting me. *False*
- No one understands me in this world. *False*
- I'm a terrible person. *False*

- Bad things always happen to me. *False*
- I am so unlucky. *False*
- This will be an absolute disaster. *False*
- Nobody could ever love someone like me. *False*
- I ruined everything. *False*
- I'm too fat. *False*
- I'm selfish and uncaring. *False*
- I can't do anything right. *False*
- I'm too old. *False*
- It's too late for me now. *False*
- I ruined the evening. *False*
- I failed my exam. *True*
- A friend in need said "no" to me. *True*
- I made a mistake and caused us to miss the movie. *True*

#5. The Socratic method

The Socratic method, sometimes referred to as Socratic questioning, involves a disciplined and thoughtful dialogue between you and your therapist or by yourself (if you can be objective enough to carry it out). It has its roots in the great Greek teacher, and philosopher Socrates and the method is applied between teacher and student, coach and coachee, and mentor and mentee. In CBT, the Socratic method is used as an umbrella term for using questioning to clarify meaning and elicit emotions and consequences. It's a great tool to help one gradually create insight and explore alternative action. Instead of using the didactic approach, which emphasizes teaching, this method focuses on personal reflection and questioning.

When applying this method either on your own or with a professional, remember the intention is to thoughtfully question yourself, not engage in confrontation or judgment. It should be a guided discovery in an open interested manner to acquire insight and enlightenment. There are certain qualities that all the questions you ask should possess. They should be concise, directed and clear, open, yet with purpose, focused but tentative, and above all else neutral and free of judgment (as much as possible).

Using this method, you can thoughtfully reflect on those distorted thought patterns to figure out how realistic they are.

How to apply this to your irrational thoughts:

Identify the thought and the beliefs that are trying to run your mind, ask open questions to bring to the surface further knowledge so you can uncover assumptions, inconsistencies, contradictions, etc. Challenge the assumptions and inconsistencies that come up. See if you can identify and replace that "hot thought" that's creating the problem or, at the very least, restate it more precisely so it can stop having a strong negative grip on you.

Create a list of the go-to questions that you could ask yourself when irrational thoughts threaten to hijack your mind and emotions. Here's a list to help get you started.

1. What is the meaning behind that "hot thought"? Why have I attached that meaning and those accompanying feelings to that thought?
2. What assumptions am I making here?

3. Is there a different point of view I could have?

4. Is there evidence to validate that what I am thinking, and feeling is real?

5. Are there alternative viewpoints to consider?

6. What are the long-term implications of feeling and thinking this hot thought?

7. What would be a better thought to have? Is there a better question I could ask myself to understand why I am having this thought?

If you're at a loss on how to come up with the best questions for your particular case, then a good rule of thumb is to use the 5W's and an H. What happened? Who is involved? When did it happen? Where did it happen? Why did it happen? How did it happen?

6

THE LINK BETWEEN PROCRASTINATION AND MENTAL HEALTH & HOW TO OVERCOME IT

M ost people are quick to judge procrastination as a sign that one is just plain lazy but is that really accurate? I personally disagree. In fact, I find myself working long and hard just to beat a deadline after procrastinating for weeks. Even back in my college years, I would put things on hold till the 11th hour when a paper was due. Then I'd pull an all-nighter and get it done. Did I enjoy that pressure and anxiety? Not really. But eventually, I got the job done. Surely procrastination isn't about being lazy because I doubt a lazy person could have the stamina of putting in such long hours and hard work. So, what is the real reason, so many of us procrastinate, and is there a correlation between this habit and mental health? Finally, can we overcome procrastination?

FACING THE CONSEQUENCES OF PROCRASTINATION

Whether you justify why you procrastinate or hate it entirely, the fact is there are always negative consequences of procrastinating on anything. Feelings of anxiety, stress, fatigue, and disappointment are typical experiences for procrastinators. There's also that dreaded feeling of looming failure and the fear that something might go wrong, and it might not be possible to fix the issue because there wouldn't be enough time anyway. I mean, imagine waiting till the last minute to complete a work project only to have the computer crash hours before you're supposed to present it to your boss, thereby losing everything. That's cause for severe stress and anxiety. According to a 2009 Ferrari, Barnes & Steel research, nearly 25% of adults living in the United States and other countries are classified as chronic procrastinators.

Procrastination is actually a condition, and it has negative impacts on our mental and physical health. In a 2007 study published in the Psychology Bulletin, psychologist Piers Steel defined procrastination as "self-regulatory failure leading to poor performance and reduced well-being." In other words, it's a form of self-sabotage. Several studies show just how debilitating procrastination can become.

One such study in 2010 titled "I'll Go to Therapy Eventually" found that procrastination and stress are interconnected. It also linked poor mental health to procrastination. There's also growing evidence that procrastination affects physical health and is a factor that can lead to hypertension and cardiovascular disease. (2015 study by Fushia M.

Sirois). Of course, that naturally leads to poor performance both at work and school. If your mental and physical health is jeopardized, how can you possibly become a high performer? Procrastinators tend to earn less, spend shorter periods of time in any given job, and hold positions with lower intrinsic value. This condition is also frequently linked to poor financial decision making in adults and low grades in students.

Despite all these negative consequences, we need to realize that it is a condition, not just about poor time management or being lazy. Procrastination for many of us can be traced back to some underlying psychological reasons. Most of the time, our reasons for procrastinating and avoiding things are rooted in fear and anxiety about failure or underperformance. Sometimes it's because we are afraid of looking stupid or being judged harshly by others. Low self-esteem and the belief that "I'm not good enough to do this" can amplify this procrastination condition. Whatever the reason might be for you, it's good to address it because being a procrastinator makes you your own worst enemy and only heightens any underlying mental issues. In cases where procrastination is a symptom of underlying mental health problems, you will need some professional help. That's where CBT is useful because, for the most part, it can help you understand and recognize your unhelpful thought patterns and behavior, which contribute to issues like depression. And, of course, most people suffering from depression tend to be professional procrastinators. Through CBT and recognizing that you need to work on the underlying problem, you can start creating structures that help you take on procrastination and finally beat it.

THE ONLY METHOD YOU NEED TO OVERCOME PROCRASTINATION

The only way to overcome procrastination is to understand that your brain is complex, and it can be your best friend or your worst enemy. There are specific chemical releases within the brain that either make it easier or more challenging for you to beat procrastination, depending on whether these chemicals are naturally released in ample quantity or not. One such crucial chemical hormone is dopamine.

What is dopamine?

Dopamine is a type of neurotransmitter that's naturally produced by your brain. Your nervous system uses it to send messages between nerve cells, which is why it's sometimes referred to as a chemical messenger. Perhaps the most important thing for you to know about dopamine is that it's your "feel good" neurotransmitter, and it plays a huge role in how you feel, think, and plan out things. It is involved in neurological and psychological functioning, which includes your mood and ability to make decisions.

Too much or too little dopamine means your ability to focus, find life exciting, and even the lens through which you interpret life will be thrown out of balance.

Aside from making you feel good, dopamine is also involved in blood flow, digestion, heart and kidney function, sleep, pleasure and reward-seeking behavior, stress response, executive functioning, memory, and focus.

How does dopamine function in the brain?

From infancy, dopamine plays a huge role in your development, and in fact, research links various mental disabilities to low levels of dopamine. Genetic conditions like congenital hypothyroidism are said to be linked to insufficient dopamine. Even Alzheimer's, depressive disorders, binge-eating, addiction, gambling, and so much more have recently been associated with dopamine deficiency.

Scientists who study neurobiological and psychiatric disorders have been interested in uncovering how dopamine works and how imbalanced levels (either too much or too little) can affect behavior and lead to disability. For the sake of our particular quest of healing from mental health issues, we are going to focus on how you can increase dopamine as a method to overcome procrastination. Here are tips for implementing immediately.

- **Get plenty of rested sleep.** Proper sleep is mandatory for your mental well-being, and it also fuels dopamine production. If you have nights of restless sleep, insomnia, or poor-quality sleep, it's essential to get help because the better you sleep, the more dopamine you will naturally produce.
- **Nutrition.** Eat foods rich in tyrosine, such as meat, fish, cheese, nuts, beans, lentils, soy, and dairy, among others. On that same note of watching your nutrition, I encourage you to avoid processed foods, high fats, sugar, and caffeine.
- **Body Movement.** Exercising daily in the way you most enjoy is an excellent hack for increasing dopamine in your brain.

- **Consider using natural nootropics**, including L-Tyrosine and L-theanine.

With the right amount of dopamine, you will feel more alert, motivated, happy, focused, and almost euphoric, which are all necessary feelings for overcoming procrastination and getting things done. The critical takeaway here is that dopamine increase can help you beat procrastination because it serves many vital neurological and cognitive functions. The effects on mood and pleasure, as well as the motivation-reward-reinforcement cycle that is created when dopamine is released in your brain, is all you need to stop procrastinating and start engaging more in your projects, whether at school or work.

TIME MANAGEMENT STRATEGIES THAT WILL HELP YOU ALONG THE WAY

Aside from increasing dopamine levels in your brain (naturally and healthily), you also need to get better at time management. This is a huge topic. There are countless books written on productivity and time management, yet it still remains a huge obstacle for most people. So, let me share a few simple time management hacks that have been working for me in recent years.

#1. Block out time for critical time-sensitive projects. In other words, schedule time to boldly face this thing that's giving you so much apprehension. By creating a chunk of time where you sit with your project, assignment, or problem, you can intelligently organize yourself and figure out how to apply the second tip.

#2. Salami Slice your tasks. This essentially means you need to chunk it down into a bitesize task so you can have mini-milestones on your way to completing the entire thing. By breaking things down into something small and manageable, the overwhelm immediately dissipates. During the depths of despair as I was going through my depression, I would break things down into small 5-minute tasks. So, all I had to do was focus on that small five-minute task. That felt manageable for me then as I got better, I added more time.

#3. Get the most unpleasant tasks out of the way as early in the day as possible. I know it might seem like it's best to start with something easy but trust me, the best time to beat procrastinating is earlier in the day. The thing that you fear most or overwhelms you the most and thus stirs up procrastinating is what you should do first thing in the morning.

#4. Don't obsess about time management. Instead, focus on managing your activities and your energy. For most of us, this will be the best-kept secret of managing time and getting things done. You need to realize that what fails us is our energy, focus, and motivation levels, not that time isn't enough (although we might think that). So instead of fussing over getting more time (which is impossible as there are only 24 hours in a day), focus on increasing your dopamine and stamina. That way, you'll stay engaged in the activity for the allocated time and actually produce something good at the end.

#5. Use the Pomodoro technique (25min sessions) to give yourself breaks in between the task, so it doesn't become overwhelming. Once you can consistently work on a task for 5 minutes at a time without losing focus, you can advance to this Pomodoro technique. It's used by leaders all over the world. Both students and professionals can develop this habit of chunking down into 25-minutes of work and 3-5 minutes of rest time. Do two or three Pomodoros at a time, then take a longer break to reward yourself for the excellent work put in. During the Pomodoro, however, there should be no distractions.

OTHER STRATEGIES THAT CAN HELP YOU BEAT PROCRASTINATION

#1. Apply the decatastrophizing technique

We learned this in a previous chapter, so you should now recognize that catastrophizing something is an irrational thought and often accompanies procrastination. For example, has there been a time where you made a huge deal out of something (an upcoming exam, a research paper, a project, etc.), and you thought, "*Oh, this is going to be painfully tough.*"? In that belief of making it seem unbearable, your brain automatically rejected the idea of taking action. Even if you were right about the fact that it would be challenging, arduous, uncomfortable, or boring, that still should trigger procrastination unless it turns into irrational thought.

So, follow the technique you learned of decatastrophizing and keep things in proper perspective. For example, you might counter that same irrational thought with "*sure, this isn't my favorite task, but I can get through it, and the sooner I do, the better I'll feel.*"

#2. Get an accountability partner.

I have found this beneficial because by asking someone to keep me accountable and cheer me on, I stay focused long enough to see it through. If you're at work, you could ask a close colleague or your boss (if you get along) to become an accountability partner. If in school, it could be a classmate, friend, or your favorite teacher. If none of those options are viable, consider joining a support group, hiring a coach, or even a therapist. Not wanting to go back or break your word to someone who matters to you is an excellent incentive to squash procrastination.

#3. Optimize your environment

The environment you work in matters a lot. A dimly lit and dull workspace for someone with mental health problems is perfect for - getting more depressed and producing nothing! An environment with lots of distractions for someone with attention deficit disorders or a person who loves being on social media is the perfect way to promote procrastination. In both these scenarios, we end up with the massive problem of low production and poor performance. Therefore, I encourage you to be extra thoughtful about your workspace. Whether you work from home or in a big office, make sure the area you do your best work is off-limits when it comes to distractions such as cellphones, magazines, and so on. Install proper lighting and create a warm, bright, and inviting ambiance. Even if you're a student, you can always ensure your desk and workspace is neat, clutter-free, and near a window or in a position that makes you feel good.

#4. Develop a reward system for your efforts

In other words, always give yourself a treat when you accomplish something that triggered feelings of procrastination. Each time you manage to beat it, reward yourself for that good behavior so you can release more dopamine and reinforce that action-oriented behavior. The reward can vary depending on how big the task was and how much procrastination you had to squash. See if you can come up with a list of things (big treats and small treats) that are healthy and make you feel good. For example, if you like flowers, buy yourself some flowers when you accomplish something. Take yourself out for a nice dinner or give yourself an ice-cream treat!

PROCRASTINATION GONE: REAPING THE BENEFITS

There are so many benefits of overcoming procrastination, and the more you do it, the more you'll enjoy them. Remember, your mental and physical health are all impacted by this habit, so the more you defeat it, the better you'll feel. It will also increase your sense of accomplishment, sense of confidence, and self-esteem. When we procrastinate, we end up disappointing ourselves because, deep down, we know it's not possible to perform at our highest and best if we are always in a mad rush to meet a deadline. You're here because you want to transform your life and create a better future for yourself. The best way to secure that better, brighter future is to eliminate unhealthy habits such as procrastination. Keep reminding yourself "why" you're choosing to make this change and keep your eyes on the new life you wish to experience. The more you stay focused on unleashing that new, healthier, better version of yourself, the easier it will be to leave procrastination behind.

7

BREAKING AND BUILDING HABITS THAT SUPPORTS A HEALTHY LIFESTYLE

Automatic habits and behaviors drive nearly half of your daily life. Did you know that? Well, consider this. You probably have a morning routine that you do almost without thinking. You wake up, brush your teeth, shower, dress, etc., without giving any conscious thought to the process. And if you look back at your routine this morning, yesterday, last week, and even last year, you'll notice it's pretty much the same. Right? We are creatures of habit because it's the most efficient way for our brain to get through the day. And while habits and routines can be useful, they can also make it feel impossible to overcome a mental health issue. The condition's persistence shows certain habits and behavior have become your default operational setting, fostering the disorder. So, the path to recovery (whether you have a therapist and an on-going program or you're self-healing) is to commit to changing some of the habits that contribute to your current situation. But we don't just want to give up bad habits without replacing them with new ones, so this one will be a two-step process that happens simultaneously. It's easy to recognize the things that harm your physical health. For example, binge-eating McDonalds and KFC will drive up your cholesterol, increase your waistline and probably even give you heart disease. But there are habits you might have that make you easily relapse back into depression, anxiety, etc.

BAD HABITS THAT YOU SHOULD CHANGE

Guilt

This is a habit, yes, and a very dangerous one. Though unseen to others, your learned behavior (perhaps since childhood) feeds your mental health problems. If left unchecked, you may find yourself in a state of perpetual guilt that prevents you from leading a healthy, happy lifestyle.

If you want to know whether this is a problem, do a thought recording exercise for twenty-one days. Notice how much you magnify issues and how often you claim responsibility for creating problems with little or nothing to do with you. Are you usually blaming yourself for things, perceiving yourself as a bad person, and struggle with self-forgiveness? Then you likely have guilt as a habit, and that thing has to be changed.

Lack of proper exercise

This isn't about getting a gym membership. It's about developing the habit of daily performing some form of exercise to get your brain and body moving. The challenge is when we feel at our worst (depressed, anxious, suicidal), that's when we need the benefits of exercise the most. Yet, it's so hard to get it done.

Regular exercise is proven to ease depression by releasing endorphins and other "feel good" hormones. It not only enhances your mood and immune system but as well your sense of confidence. There's no particular form of exercise that is mandatory, but you will need to make an effort and start doing something. Consider trying different things at home, in the gym, or dance studio until you find something you enjoy doing at least six times each week. You can experiment with Yoga, Pilates, Strength Training, High-Intensity Interval Training, Dance, etc.

The worst thing you can do when overcoming a mental health issue is to exercise irregularly or not at all.

Poor quality sleep

Many people with mental health issues spend a lot of time in bed supposedly sleeping, but it's usually very crappy sleep. On the other extreme, there are those who barely get any because of insomnia. In both these cases, insufficient sleep is a massive struggle during recovery and treatment. It should be addressed as early on in the treatment as possible because it's hard to heal without a night of good restful sleep. The Sleep Health Foundation reports that 60 to 90 percent of patients with depression also have insomnia. According to the Harvard Mental Health Letter published by Harvard Medical School, poor sleep can result in mental health problems, and treating sleep disorders can relieve mental health symptoms. So, if you weren't taking this seriously, it's about time you start.

A CHALLENGE TO SELF: 3 STRATEGIES TO BREAK THE BAD HABITS

Now that you have an idea of some of the habits that need to change in your life let's talk about how to break them. I want you to challenge yourself with these three strategies:

First, figure out your triggers

For example, if your habit is procrastination or stress eating, pay attention to the circumstances surrounding you when you do those things. Once you identify your triggers, take notice of the impulsive behaviors that follow as you act out. So, if it's procrastinating, do you reach for your phone instead and drown in social media instead of starting the project? When someone stresses you, do you immediately go for junk food or something sweet as you sob and worry?

The second thing you must do is:

Increase your awareness and become mindful of your feelings as you're doing this

It's vital to employ mindfulness at this point so that you don't attack yourself once you catch yourself in the act of a bad habit yet again. As you start doing "the thing," pay attention to how you feel when partaking in it. If you're stress eating and down to your second box of cookies even though you're not hungry, just pause for a moment and get in tune with how you're really feeling. What is your state of mind right now? Do you really feel good? How good will your mind and body feel after another hour of getting through those boxes of cookies? Is this really the kind of lifestyle and behavior that makes you happiest? Grab your journal and just write what you're feeling deep in your heart as you ask these questions and self-reflect.

As your awareness increases, your brain will accurately update the reward value of the habit you want to break, and it will begin to see that "X" behavior leads to "Y" consequences, which isn't what truly makes me happy.

The third thing you must do is:

Let go of your failure mindset

It's hard to be perky and enthusiastic when struggling with depression, anxiety, and so on, but your healing can only work when you let go of that doom and gloom mindset. Those negative thoughts are actually a habit that can be released, but they've been around so long you probably assume that's the only way to think. Ugly irrational thoughts will tell you that this won't work or that there's no hope for you, but you can and must choose to think differently. This will work this time, and you do not need to be tortured by your own thoughts of failure anymore. As you learn to disregard these thoughts and allow them to float away from when, they came, you'll find the courage and optimism needed to continue on this quest of breaking the habits that have caused you so much psychological pain and suffering.

BUILDING ON THE GOOD HABITS

Breaking bad habits isn't enough. You also need to build the right ones to replace the old, outdated ones that no longer serve you. But how does one start doing this? I recommend starting simple. Take small steps forward and implement one habit change at a time. Don't do too much too soon. Here's a scientifically proven technique that will help you build new habits.

THE HABIT STACKING AND THE ANCHORING TECHNIQUE

Habit stacking is a method created by Dr. BJ Fogg as part of his Tiny Habits Program. It's about building new habits by stacking them on top of a current one that already exists. The formula is quite simple:

After/Before [current habit] I will [new habit].

When it comes to helping your brain form new habits, the best thing to do is leverage existing habits that are working well. Ever noticed how efficient you are at remembering to take a shower and open the blinds each morning as you start the day? Or how you automatically know where to throw the keys as soon as you walk into the house? It's not something you have to consciously plan. Your brain has built a strong network of neurons to support that behavior of getting out of bed, drawing up the curtains and getting into the shower or opening the door and immediately offloading your keys and coins, placing them in the same location each time. The more you do something, the stronger and more efficient the connection becomes. So how would this apply in the creation of a new habit?

Instead of pairing your new habit with a particular time and location, pair it with a current routine.

Examples of implementing habit stacking:

After I sit down to dinner, I will say one thing I'm grateful for that happened today.

Before I pour myself my morning cup of tea, I will meditate for five minutes.

Overall, habit stacking allows you to intentionally implement new actions with will be ingrained in your brain more quickly because you are stacking them into something you're currently doing. It's like creating a game plan for which action should come next. Once you get comfortable with this approach, you can develop general habit stacks to guide you in all situations. For example, when I walk into a party, I will introduce myself to anyone I don't know yet, or when the phone rings, I will take a deep breath and smile before answering.

GOOD HABITS YOU'D LIKE TO HAVE

#1. Work on your posture

Most of us don't realize how much poor posture affects how we feel. Evidence shows poor posture can cause one to feel more depressed, anxious, and insecure. Therefore, it's time to correct your posture. Start by standing up and sitting up straight. Notice the difference it makes even if you're just practicing in front of the mirror for a few minutes each day until it becomes the new norm.

#2. Drinking plenty of water

Hydration can mess with your focus and mood because the brain requires a lot of water to function optimally. Research also shows that water boosts mood and energy levels. Most of us forget to hydrate enough, so a good habit to develop is drinking at least 10-12 glasses of water daily. Consider using the habit stacking and anchoring technique in the morning for drinking 2-3 glasses of water as part of your morning routine.

#3. Spend some time with the morning sun

The morning sun is excellent for aiding your body in the synthesis of vitamin D. In fact, something I noticed through sheer experiment was that five minutes with my face to the sun early in the morning significantly reduced my depressed state. Scientists seem to agree that not enough sunlight can cause depression, which is ironic because depressed people hardly ever want to be outdoors. But you don't need to leave your house, just stand next to an open window and stick your head out for a few minutes or sit on your terrace each morning. It will give you an energy and mood boost instantaneously, enabling you to tackle the day better.

#4. Read something inspiring

One of the best habits you can pick up is starting and ending the day with something inspiring, motivating, and uplifting. Reading the right materials also broadens your thinking and increases your mental power. Consider reading self-help books or even getting crosswords, jigsaw puzzles, and other mentally stimulating activities to keep your mind active.

#5. 4:55 Drill

This is a little technique that can massively improve your productivity and beat procrastination if you turn it into a habit. At the end of each workday, use the last five minutes of your day to get yourself organized for the following day. Take a moment to decide on no more than two things you'd like to accomplish first thing in the morning and assemble everything you need now so that you can hit the ground running the next day. All it takes is the last five minutes (hence 4:55 Drill), but this small shift in how you end your day can significantly help you eliminate that feeling of overwhelm or procrastination in the morning.

#6. Journal daily.

The recovery process is long, and there's nothing more therapeutic than getting into the habit of journaling your thoughts at the end of the day so you can express how you're feeling as you go through this recovery process. It's also a very effective way of monitoring your self-talk, which leads me to my next suggestion.

#7. Be self-aware and practice positive self-talk

As you practice mindfulness and increase self-awareness, you can begin to be mindful of your feelings and emotional responses. Monitor the inner dialogue you currently have and make it a habit to speak more kindly to yourself. As you speak more positively to yourself, you'll find it easier to also watch your tone when interacting with others, which improves your social skills.

#8. Clean up your environment

This is a simple action that can create a sense of ease and lighten your mood, making you feel more ready to face the day. A clean workspace, clean home, and clean bedroom are all great for promoting a sense of well-being and organization. I don't know about you, but when I am going through depression, the last thing I want to do is make my bed or clean my office space, yet these cluttered environments do very little to help with recovery. Research shows that your personal environment directly impacts your well-being. A clean home and work environment can influence your mood, affect your behavior and motivate you to take action. It also helps reduce that sense of constriction or stagnation and stress because a clutter-free area has better air circulation, less stuffiness, and a sense of expansiveness, which will impact how you feel internally.

#9. Practice gratitude daily

Invest five minutes a day to write out things you are grateful for. I started this as a morning habit, and it has helped me curb those negative thoughts that used to hijack me in the morning. I find some sunny spot in my house, sit with a pen and paper and jot down five things I'm grateful for. Some days are easier than others, but even when I feel like I have nothing good going for me, I still force myself to feel grateful for the things we often take for granted, like the fact that I had running water in my house or that I took a hot shower. Try it for one month, and you'll see the therapeutic benefits this has.

SURE-FIRE METHOD TO MAKE THE NEW HABITS STICK

Building a new habit isn't easy and doesn't happen overnight. You also can't expect to see immediate results even once you start implementing the action and behavior, so how do you ensure you stick with it long enough to reap positive rewards?

It begins with a shift in mindset. What you need to focus on is making this a part of your new lifestyle. It's not about whether something is giving you the end results you need. It's about whether it's feeling naturally integrated into your life and whether you're enjoying it as part of your journey. For example, if you decided to eliminate the bad habit of no exercise with a morning jog, your goal shouldn't be to lose weight in a week or even two. Implement the new habit and train yourself into waking up each morning and doing your morning run simply because you want it to become the new way of living. While running, you also catch that morning sun, and you're outdoors in nature, allowing you to habit stack multiple good habits all at once. The end result of losing weight and getting fitter should be a by-product that comes in its own due time. Your reward should be more centered on how you feel during and after investing in this healthy habit.

Another sure way to help you maintain your new habits is to keep yourself accountable through a habit tracker. A habit tracker is a simple way to measure whether you implemented your habit or not each day. Research has shown that people who track their progress on goals like losing weight, quitting smoking, etc., are more likely to improve than those who don't. So why not set yourself a goal of positive self-talk, drinking enough water, jogging in the morning, or making your bed first thing in the morning?

You can get an app on your smartphone or use a calendar and simply cross off each day you stuck to your routine. What you'll notice over time is that your habit will start creating streaks that you'll feel compelled not to break. That reinforces that feeling of winning and healing that's extremely desirable. This is one of the best signals and proof that you're making progress. It motivates you to keep going and feels very satisfying as you see yourself accomplishing something productive each day.

There are many options for habit trackers on iOS and Android, or simply get a journal with a calendar and do it manually. Regardless of the format, you use to track the new habit, keep it simple and easy to do. Also, remember to start small by integrating a single habit at a time until it becomes part of your daily life.

8

MINDFULNESS: WHAT IS IT AND HOW CAN IT HELP YOU

Y ou've likely seen a lot of mindfulness talk on the Internet. It seems to be the trend that many are promoting as a tool for healthy living, from celebrities to athletes and everyday busy individuals. Most of what people talk about when referring to mindfulness practices is meditation. Although meditation is a powerful tool, it's not the only way to practice mindfulness. So, if you tried it and didn't reap the benefits, don't give up just yet. I encourage you to keep reading so you can discover more ways to practice mindfulness.

A LOOK INTO MINDFULNESS

To understand mindfulness, we need to define it. What is mindfulness? According to American Psychological Association (APA.org, 2012), mindfulness is a moment-to-moment awareness of one's experience without judgment. In this sense, mindfulness is a state and not a trait. Merriam-Webster Dictionary also offers a unique under-standing of mindfulness as it defines it as the practice of maintaining a non-judgmental state of heightened or complete awareness of one's thoughts, emotions, or experiences on a moment-to-moment basis. One of my favorite definitions is from Greater Good Science at the University of California at Berkley. "Mindfulness means maintaining a moment-by-moment awareness of our thoughts, feelings, bodily sensations, and the surrounding environment." While there are varying definitions, the fact is, mindfulness is about present moment awareness. When it comes to mental health problems, mindfulness is extremely helpful because it keeps you focused on the here and now instead of consumed by the past or the future. Often, our biggest problems arise from spending too much time in the past or in the future.

WHERE DID MINDFULNESS COME FROM?

Mindfulness isn't just a buzzword. It's a practice that has been around for a really long time. However, in the West, Jon Kabat-Zinn is linked to this concept of mindfulness because he re-imagined Buddhist contemplation practices for the secular world decades ago. Since mindfulness is centered around knowing the mind, training the mind, and freeing the mind when used as part of the healing treatment, it can be an effective way of healing and creating your dream lifestyle.

Two components work together to bring about relief. Awareness and an open and accepting attitude. Both are essential when you start practicing mindfulness. Research shows that people who practice mindfulness receive heightened metacognitive awareness. That means one can detach from one's own feelings and mental process. It also decreased patterns of negative thinking behavior, which positively affects the individual and reduces the chances of a relapse into depression after treatment is completed. Other studies conducted also showed that mindfulness can improve anxiety and depression symptoms and reduce stress levels. In short, anyone struggling with mental health issues can significantly benefit from this practice.

CAN IT REALLY HELP ME?

If you struggle with anxiety, depression, and other similar mental health issues, then mindfulness is definitely for you. Anyone can learn and practice mindfulness whether you have severe mental disorders or simply want to improve your current mental state. The best part is anyone at any age can implement mindfulness into their daily routine. Choosing to practice mindfulness will increase your awareness and help you detach from irrational thoughts and overwhelming emotions. It will help you center your mind, improve your memory and bring more clarity into your life.

It will make you more accepting of yourself and situations that are beyond your control. Instead of resisting or fighting your fears, doubts, or even anger when it shows up, you can simply learn to observe and let it go.

MINDFULNESS AS A THERAPY

It is possible to combine mindfulness and therapy. In fact, most research centers around two specific types of mindfulness training.

The first was pioneered by Jon Kabat-Zinn, known as mindfulness-based stress reduction (MBSR). The second is Mindfulness-based cognitive therapy (MBCT), a type of psychotherapy that combines cognitive therapy, meditation, and the cultivation of a present-oriented, non-judgmental attitude, aka mindfulness. It builds upon cognitive therapy principles by using techniques such as mindfulness meditation to teach people to consciously pay attention to their thoughts and feelings without placing any judgments upon them. This type of mindfulness training was created by a group of therapists (John Teasdale, Zindek Segal, and Mark Williams) who felt the need to develop a cost-effective method for treating and preventing relapse in depressive patients.

When treating chronic depression, the goal of MBCT is to help a patient learn how to avoid relapses by not engaging in those automatic thought patterns that often worsen depression. A recent study showed that MBCT reduces the risk for relapse by 50% regardless of age, sex, education, or relationship status.

HOW DOES IT WORK?

An MBCT program is usually a group intervention that lasts eight weeks. In the program, you attend a weekly course that lasts two hours and one day-long class after the fifth week. During this time, you'll be taught what's known as the three-minute breathing space technique, which focuses on three steps, each one minute in duration. Step one is observing your experience and how you are doing right now. Step two focuses on your breath, and step three is attending to the body and physical sensations.

When combining mindfulness and therapy, most of the work is self-directed. So even if you choose to combine mindfulness with Cognitive Behavioral Therapy, Dialectical Behavioral Therapy, or Acceptance and Commitment Therapy, you still need to make that effort to become more aware of your thoughts, feelings, and actions.

HARNESSING THAT INCREDIBLE POWER

Now that you've been introduced to the power, becoming aware, and keeping your focus in the present moment, it's time to start putting it into practice so you can harness that incredible power. You can train your mind to be more "in the present moment," freeing yourself from worry, trauma, and anxiety. By activating this power, you become more equipped to face and make peace with whatever challenges life throws you. It all begins with your willingness and decision to spend more time being present. As you eat, walk, interact with others, work, and so on, you can keep giving yourself little reminders to stay focused in the Now. You can also read books to help you understand the importance of present moment awareness, such as "The Power of Now" by Eckart Tolle. You can also learn from the great master himself, Thich Nhat Hanh, who has several books and YouTube videos on making mindfulness part of everything you do. As you train and still your mind, you will discover an aspect of you that has been thus far missing from your life. Mindfulness is a sure way to transform your life. Begin today.

MAKING A HABIT OUT OF MINDFULNESS

Let's help make this concept of mindfulness practices with simple step-by-step tips on applying it in the things you're already doing.

Tip #1: Do one thing at a time.

Forget what you hear people say. Multi-tasking will get you nowhere. It's not a productivity hack. If anything, it just spreads your attention and energy really thin, making it even harder to accomplish tasks. Therefore, give yourself permission to slow down and focus on one activity at a time. Take on each task with your full attention and bring awareness to what you're doing and how that feels. This is a simple way of increasing your power of concentration and mindfulness. In so doing, you are less prone to rush, forget details or make silly mistakes. You'll notice over time that you move through your activities with greater ease and confidence because you give yourself enough time and personal presence.

Tip #2: Sit in silence and observe your mind's chatter.

Anytime you watch your thoughts, you're actually being mindful. You can invest a few minutes during the day to listen to the voice in your head without self-loathing or criticism. Notice when repetitive thoughts enter your mind. Be an observer, not a judge and what you'll start to notice is that there are voices in your head and some thoughts are disconcerting... but... you are not your thoughts. Do this long enough, and you'll come to the full awareness that you are not your mind, and so you can choose to detach from the noisy playground called your mind.

Tip #3: Mindful walking.

This is one of my favorite things to do because I'm not a fan of the classic meditation in the lotus position. Instead, I find walking in the park or anywhere in nature to be more beneficial in restoring my sense of calm. Whether you're walking to work, around your neighborhood, from the car to the store, or through the hallways at work, you can choose to turn it into a meditative experience.

How? Well, try this.

The next time you're about to open the car door or rise out of your chair at the office, turn your attention to your intention of walking mindfully. Simply say to yourself, "I'm going to be fully present and mindful of each step I take." Then make your move. As you do, become aware of the sensations. Put your attention on your body. Pause, take one conscious breath. Begin to move your feet. If time permits, you can deliberately take slow steps to feel the moment more and more. Notice how the ground feels under your feet. Notice how your clothes feel against your body as you walk. If you're doing this outside, notice the air, birds, trees, plants, and all the other little details you would often ignore. The goal here is to be present with each step for as long as you can.

Tip #4: Mindfulness listening.

This is such a great and powerful habit to develop. Active listening through the lens of compassion will significantly enhance your relationship with yourself and with others. Most of us don't realize how disconnected we are from our bodies and feelings. And if we are disconnected from our own bodies, we can't possibly be present enough to connect with another. So, what you might realize (like I did) is that I am usually caught up in my own mind chatter as others speak to me. Here's how you can shift to being a mindful listener. Practice noticing your own thoughts as shared in the first tip. Next, make an effort when a loved one speaks to you and just listen to what they are saying. Don't carry on an inner argument or dialogue. Focus all your attention on that person. You'll be amazed at how different that interaction will be. People tend to unconsciously pick up whether we are present or not.

Tip #5: Do a body scan.

The body scan is another way to bring mindfulness into your daily routine. It can be done in a matter of minutes or as long as half an hour, depending on your lifestyle and needs. You can also choose to do it as a morning or evening ritual. It begins with you lying on your back with your palms facing up and feet slightly apart. If you're doing it outside your home, you can also do it sitting on a comfortable chair with feet resting on the floor.

Once you've settled into a comfy position, immobilize your body and then begin bringing awareness to your breath. Notice the rhythm, the experience of breathing in and exhaling out. Do not control or manipulate the breath but simply observe.

Next, I want you to notice your body: how it feels, the texture of clothing against your skin, the contours of the surface on which your body is resting, the temperature of your body, and the environment. Are you noticing any tingling, soreness, etc.? What sensations are you aware of now? Does your body feel particularly light or heavy?

Are there areas that feel hypersensitive or areas where you feel numb?

Notice that a typical body scan runs through each part of the body, paying particular attention to the way each area feels. It's advised that you move from your feet to the top of your head, i.e., toes of both feet, the ankles, lower legs, knees, thighs, pelvic region, abdomen, chest, lower back, upper back, hands, arms, neck, face and head covering as much detailed ground as you can.

Once you complete the body scan, bring your awareness back into the room and slowly open your eyes. Many videos on YouTube can also guide you through a meditative body scan with ambiance music.

A SIMPLE 5 MINUTE DAILY ACTIVITY THAT BOOSTS YOUR MENTAL HEALTH

One of the most promoted mindfulness tools is meditation. Although it doesn't work for everyone, research proves that meditation is an incredibly powerful way to boost the immune system, reduce stress, improve mental focus and clarity, among other things. It also helps curb negative self-talk and the tendency to fall into irrational thoughts. If you've tried it before and struggled to stick with it long enough, experts say it might be because you're doing it too long or focusing on the wrong objective. As a beginner, as little as one minute of mindfulness meditation can be sufficient enough to get you going. Ideally, experts recommend five minutes of mindfulness meditation. Why? Because five minutes is considered ample time to familiarize yourself with the simple act of sitting in stillness in the midst of a chaotic day or a racing mind. So, you can do this five-minute meditation in the morning, during your lunch break, pre-bedtime as you wind down, or just before sleeping.

Here's what to do:

First, you need to set a gentle five-minute timer then find a relaxed, comfortable position that works for you. You could sit in a lotus position or sit on a chair with your feet on the ground. You could also sit on the floor on a cushion. Keep your back upright but not too tight and rest your hands wherever feels comfortable. You can keep your tongue on the roof of your mouth if that feels comfortable.

The second is to bring your attention and awareness to your body. Try to notice the shape of your body, how light or heavy it feels, and just let yourself relax. Notice the sensations and relax any tension or tightness. Just breathe in and out naturally.

The third is to tune into your breath more consciously. Feel the natural flow of your breath without altering anything. Notice where it's easiest to connect with the breath. It can be your chest, throat, nostrils, or abdomen. There's no right or wrong here. It's all about connecting with yourself and remaining at ease as you focus one breath at a time.

The fourth thing is to practice compassion and kindness to yourself as you catch your mind wandering. You'll notice very often that you got distracted and are no longer focusing on breathing in and out. That's okay. As thoughts come and carry you away to something that happened in the past or what you think will happen in the future, be kind to yourself. This is natural, and we all go through this. Softly and gently redirect your attention back to the breathing.

Keep noticing your breath in silence for the entire five minutes until the timer goes. Don't worry about getting lost in thought; just keep returning to your breath. Once the gentle alarm goes off, start coming out of the meditative state and check in with your body again before bringing back your attention to the current environment.

Give yourself just one more minute to think of a positively inspiring thought such as "May I feel more grounded and calm today," and appreciate yourself for participating in this mindfulness experience and for the gift of being alive and having that breath. Now you can get up and carry on with the next activity.

THE GUIDED MEDITATION SCRIPT

I will now take you on a journey of relaxation, visual imagery, and pure visualization. You will learn to leave your problems and inner anxieties behind and will gain a new understanding and clarity of mind. Embracing instead a powerful and vibrant visualization that fills your being with wonder enabling you to understand your place within the world and all that is important. You will learn to let go of tension and impress into your mind powerful positive statements that will improve your sense of well-being.

Before we begin, please ensure you are sitting or lying down in a comfortable, well-ventilated room where you will not be disturbed for the next twenty minutes. You may read this meditation first and if it resonates with you, consider recording it with your own voice and adding some background music of a healing nature, such as Tibetan meditative music, to further enhance the experience. As you follow this guided meditation, feel free to keep your eyes closed and allow yourself to escape from the current environment and restrictions into the powerful moment presented by this meditation. Nothing else matters but the here and now. At this moment, there is nothing for you to feel concerned about. You are at peace. You will allow the tensions of the day to dissipate. You will give yourself permission to connect with the universe. Let us begin.

Close your eyes. Take a deep and slow inhale through your nose and then slowly exhale through your mouth. Repeat this a few more times, each time filling your belly completely as you breathe in and then exhale till it's completely contracted as you breathe out. As you exhale, picture any tension leaving your body as a color. Let the tension fill the airs swirling around you. If you feel angry, envision the breath as a deep red color if you can to make it vivid. Allow any and all tension to dissipate as the breath leaves your body.

Now, inhale again. Breathe in slowly through your nose for a count of four (one, two, three, four) ... Feel and see yourself breathing in a color representing peace (it can be blue, white, or any other color you like). Extend your diaphragm as you feel the air entering your lungs. Breathe in deeply to the bottom of your lungs. This time, with your lungs full of air, I want you to pause and hold the breath for a count of two (one, two) and then exhale slowly through your mouth. Inhale deeply and slowly, hold one... two... exhale slowly as you control the outflow and count one...two...three...four. Notice the color that you are inhaling and the color that you're exhaling. Continue this cycle of breathing for a few more minutes. This is rhythmic breathing. It's perfect whenever you feel tense, stressed, or nervous. Inhale calming energy, exhale and release any worries, anxieties, or physical tensions.

Inhale slow and steady to a count of four. Hold the breath for a count of two. Exhale slow and steady for a count of four.

Do this for one more minute with background music or in silence.

We carry a great deal of tension in our neck and shoulders. Raise your shoulders slowly up to your ears, hold for one...two...three, and release. Notice and acknowledge that your body is starting to feel more relaxed. Keep using your breath to relinquish any tension and feel your body begin to relax more and more.

Your arms and legs are now feeling heavier. The muscles in your back are relaxing. If there is any tension left in your shoulders, contract and tighten your muscles, hold for a count of four and then release. Feel the shoulders relax. Feel your back supported where you are now, and just enjoy the sensation of breathing and relaxing. Do this for one minute.

As you continue breathing normally without seeking to control or manipulate your body, bring your attention within. Go into your imagination and see yourself standing at the end of a hallway next to a room filled with boxes of varying sizes. Some are really small, and others are huge. This room contains boxes filled with all your problems, anxieties, worries, and regrets. Right next to you is a stone case staircase that spirals around. The steps are made from white marble. You begin to climb up the stairs supporting yourself with your fingers along the stone wall. You feel the smoothness of your stone as you touch it. It's cold and smooth. Continue to climb slowly, making your environment as vivid as you can. There are many steps spiraling up ahead of you. You feel courageous enough to keep climbing. It feels wonderful to be leaving all your problems below. And as you continue to climb, you see below you a small room. Feel a sense of relief as you climb higher, moving further away from all the clutter, chaos, and anxieties. With each step, you move towards peace and inner contentment.

You are now approaching the top step, and an inky blackness greets you. You are not afraid as you emerge onto a circular platform. You know that the sky forms the roof, and there are millions of stars twinkling in the black expanse. There is a sudden rush of freedom. Feel the sense of wonder and awe as you look toward the heavens. You are free.

In the middle of this pure, white, curved platform that you stand on is a circular flat seat slopping back into a contoured chair made out of the same marble as its surroundings. Sit and feel the coolness of the marble beneath you. It is a perfect place for reflection. Lean back and feel the stone supporting your back in all the right ways. It's as if it was carved with you in mind.

Feel how comfortable and serene it feels to be sitting here. Now, look up into the night sky. There are no clouds to mask the stars. The whole vast sky is open to you. Here you are free from the tensions of everyday life. Here you are just as perfect as the stars that are shining bright for you.

In real life, you may feel frustrated, confined, tense, and even trapped, but as you look up now into the depths of this extraordinary sky decorated with millions of stars light-years away, you feel a new sense of desire emerging within you. You feel like floating up high to meet with and become one with the universe. The desire to feel weightless, to gain a new and inspiring sense of perspective rises. You feel yourself begin to float gently out of the marble seat. You move up higher...and higher... and higher. You float up above the marble walls, and now you can see the view of the city around you. Lights that twinkle far beneath the expanse of sky. The view is magnificent. You can see for miles. The landscape lit by twinkling lights heralding the existence of those who live and share this reality with you. A cityscape that comes to life with the sprinkling of artificial lights that mirror the heavens above.

You move effortlessly, relishing in this feeling of freedom. Here you are cast afloat from the problems of daily life. It is like flying, but with no effort required. A single thought enables you to change direction at will. And you travel on. You look down. Moonlight is reflected in the rippling waters of the estuary. Gentle waves lap at the shore, and as you float out further high above the darkened waters, boats bob along the harbor walls, and the feeling here is of complete peace and tranquility. You feel invisible. There is no blame here, no regrets. Just awe at being able to see life from a whole new perspective. Traveling over the estuary, you head along the coastline scaling large cliffs, flying high above them. Thin clouds almost transparent from below are blown in from the sea, hugging the cliffs as you move higher. The clouds, wispy, fragile, and translucent, follow in your trail as you soar higher. Rising vertically now higher and higher, looking up toward the moon. Silvery grey, the moon is full. You marvel at its beauty and power, knowing how it controls the ebb and flow of the tides.

Here, suddenly, life feels less complicated. There is a sense of pureness. Of mystery and yet clarity. Life is good. Life is wonderful, and you share a sense of connection with the universe.

Breathe in deeply and then release the breath. Let go of any tension within your body. Focus on the softness of every muscle. Feel yourself free, relaxed, and free from any burden. Far below, you see low flying birds hug the surface of the water disturbed. They seem tiny and fast-moving. The water ripples gently. Sparkling stars and the light from the moon herald your way. You see the world from an enlightened viewpoint, and it is one of wonder. Here, high above the fragmented clouds, you feel a part of the world's mystery connected on every level and free from your problems. You can sense a change around you. Gradually the air becomes warmer and misty as cloud formation begins to forge together. The sky changes color. Dark muted through to light, and for a while, the sun and moon share a place in the heavens. You drop down now through the fluffy clouds and float down towards the earth's surface. You feel exalted as the sea now mirrors the changing skies—sparkling sunlight glinting across the surface and shimmers of blue. You drift down, moving on away from the water's edges still high above the city's architecture, tall concrete pillars and homes reaching up toward the sky and small private residences alone surrounded by small patches of green like a complex jigsaw of life.

Stone and irregular shapes all fitting in together in almost seamless ease. Marvel at how the man-made structures can look beautiful too. Here, you can see that life is about living and not holding onto problems. There is no place in your life anymore for anxiety that makes you feel ill. Tension headaches, depression, regrets that gnaw away at you, or the decisions that you just can't make. Here in this weightless existence, you realize that you can be free from all of the negative aspects of life, no longer shackled by an existence that holds you back. This acknowledgment is meaningful. It has the power to change your life.

As you breathe in deeply, conjure up the image of those problems and all of the boxes that you left behind. Initially, they were overwhelming, threatening in their power to hamper your life. To impact you at every turn. Now picture them reduced in size, no more threatening, just minor inconveniences that you have now scaled back in your mind. Shrink the problems smaller still. Breathe deeply again and then out. Breathing out the last of your connections to those problems and see them growing ever smaller, every minute, a fraction of their former size, and you realize your perspective was clouded before. Your judgment was off-center. Your realization of the truth in life and the importance of nature and your place in the world makes those problems seem insignificant by comparison.

Focus again on your breathing. It is time to see the problems of life as mere obstacles. It is time to look at the important things in life and to step away from any doubts, regrets, or anxieties that affect you negatively. It is time to feel contentment and to embrace pure inner peace. Drift gently now down to earth. Close your eyes and feel your descent. You are at peace with life and with yourself. Finally, you understand your part within the universe. You drift down as you're bathed in a golden light as the sun shimmers in the early morning sky. The faint warmth is comforting—a beautiful start to the day.

It is time to feel positive about your life. You have the power to do so. You are centered. You retain the feeling of peace and wonder. You are now back in your own reality. Feel yourself back in your bed, comfortable, safe, and secure. Open your eyes and stretch out your muscle. Breathe deeply to send out oxygen to your body. Realize how good you feel right now.

Keep experiencing the incredible sense of calmness and deep peace as you remember your wonderful high time above the clouds. A part of the midnight sky and then a part of the early morning transition as night became day. Understand now that your problems are such a small part of time and space and even within your own reality, under-stand that your problems can be dealt with quickly and clearly with clarity of mind. Breathe in deeply and keep your sense of peace and tranquility. Breathe out and noting that no tension remains. Remember that you can return to this meditation whenever you need a renewed perspective. You are at one with the universe. Namaste.

9

THE RELATIONSHIP BETWEEN SPIRITUALITY AND PERSONALITY

A lot can be said when it comes to mental health, spirituality, and religion. There's been increasing evidence showing benefits to having a healthy spiritual or religious life (depending on your preference and beliefs), but it wasn't always the case. In fact, up until the early 1990s when "religious or spiritual problems" were introduced in DSM-IV as a new diagnostic category that invited medical professionals to respect the patient's beliefs and rituals, psychiatrists who are generally less religious than their patients didn't value the role of spirituality or religion. Why? Because ever since the falling out of religion and psychiatry in the early 19th century (thanks to Charcot and his pupil Freud who associated religion with hysteria and neurosis), mental health care and all things religion were divided.

But there's no denying the fact that substantial evidence shows someone's spiritual inclinations do influence their mental health. So, is it worth taking into account? This is what we want to explore in this chapter.

A SPIRITUAL LIFE, SHOULD YOU HAVE ONE?

There's no right or wrong answer here. It all depends on your beliefs and world view. Spirituality or religion (they are not the same) can have a tremendous impact on your recovery if it aligns with your personality. But what is spirituality, and how is it different from religion?

Different people will offer different definitions for spirituality, but I choose to think of it as a way of thinking about the meaning and purpose of one's life. Spirituality is about finding your sense of worth and value in this life. It offers a worldview that suggests there is more to life than just what people experience on a sensory and physical level.

Spiritual practices include things like meditation and prayer, living by a set of rules that you establish for yourself (like how you treat people), and focusing on values such as kindness, compassion, hope, honesty, and equanimity. Spiritual people naturally prioritize mindfulness practices, which makes it easy to integrate therapies that use mindfulness. Religion, on the other hand, is linked to a particular faith, tradition, or institution. Being religious usually entails believing in the god of that particular faith. You have specific religious leaders who guide you through the shared and commonly accepted beliefs as you go through life. One can be spiritual without being religious, so this isn't a matter of forcing any particular faith upon you. It isn't even about forcing you to become a spiritual being but rather an invitation to get curious about whether your personality, well-being, and sense of purpose will be significantly enhanced by the contemplation and practice of spirituality.

For some people (including myself), awakening to my spiritual nature was part of the accelerant for my healing. I struggled for most of my life, falling in and out of depression and other mental disorders. I would undergo treatments and become "normal" for a while but then eventually fall right back into it. There was always a reason. The last big reason was my relationship gone wrong, and it just threw me a curveball that felt too big for me to handle. It wasn't until I surrendered my life to a higher awareness and reconnected with my spiritual self that treatments started working. For me, the benefits are too numerous to mention, beginning with the fact that I am alive and healthy, and I now have enough courage to share my insights on how to heal with you through this book. All that happened because I reconnected my personality with the spiritual aspect of myself.

Research has shown that those who are more spiritual or religious and use their spirituality to cope with life challenges experience many benefits to their health and well-being. If you are part of a spiritual community, you may have more support and friendships that empower you to stay positive as you go through treatment and recovery. It might be helpful to feel that connection to something bigger than yourself. Becoming spiritual can give you strength and hope, especially when you're feeling unwell. It might also help you to make sense of your experiences. All this and so much more is possible if you feel inclined to follow this path. But I must state that you are the only person who can make this choice. It cannot be forced. If you want to begin thinking or talking about your spiritual needs, find someone you trust or join a community of people whose values you share. But a few questions you can contemplate to figure out if spirituality can help you transform your life include:

1. What is important to you?
2. What gives you hope and keeps you going when things get tough?
3. Do you have a feeling of belonging and being valued? If no, would you like to?
4. What makes you feel supported?
5. Do you feel safe?
6. What makes you feel happy?

SPIRITUAL INTELLIGENCE: A CONCEPT

Spiritual intelligence (SI) is a concept that was introduced by Danah Zohar in her book *ReWriting the Corporate Brain* back in 1997. It's a term that's now widely used by some philosophers, psychologists, and developmental theorists to indicate spiritual parallels with Intelligence Quotient and Emotional Quotient. Famous author Stephen Covey also believes spiritual intelligence is essential, stating "it is the central and most fundamental of all the intelligences because it becomes the source of guidance for others."

Zohar speaks of 12 principles underlying spiritual intelligence.

1. **Self-awareness:** Knowing what I believe in and value and what deeply motivates me.
2. **Spontaneity**: Living in and being responsive in the moment.
3. **Being vision – and value-led:** Action from principles and deep beliefs and living accordingly.
4. **Holism:** Seeing larger patterns, relationships, and connections; having a sense of belonging.
5. **Compassion:** Having the quality of "feeling-with" and deep empathy.
6. **Celebration of diversity:** Valuing other people for their differences, not despite them.
7. **Field independence:** Standing against the crowd and having one's own convictions.
8. **Humility:** Having the sense of being a player in a larger drama of one's true place in the world.
9. **Tendency to ask fundamental "Why" questions**: Needing to understand things and to get to the bottom of them.
10. **Ability to reframe**: Standing back from a situation or problem and seeing the bigger picture of a wider context.
11. **Positive use of adversity:** Learning and growing from mistakes, setbacks, and suffering.
12. **Sense of vocation:** Feeling called upon to serve, to give something back.

The more we study spiritual intelligence, the more we see evidence of the relation between personality, spirituality, existential intelligence,

and spiritual intelligence. Your personality is of paramount importance in the process of creating the inner motivation that triggers the
innate human tendency to search for the meaning of life. In that
quest, one engages that dimension of spirituality, which then activates
existential intelligence to develop a system of beliefs and values and
the capacity to tackle deep questions about human existence, which is
usually very useful when one desires to understand the meaning of
their life. The last stage is spiritual intelligence, which helps direct the
right and necessary actions to implement the intended goals. There
need not be any religious acceptance in all this, and that's why it's
important to note that although spiritual intelligence does connect to
spirituality, it does not in any way require religious precepts.

While SI can be adapted in various settings, including the workplace
and within the context of how you relate to your life and others, it is
important for your well-being. Research conducted in the past few
years shows a correlation between SI and satisfaction with life. It is
speculated that SI creates an environment that promotes inspiring
self-reflection and prompts an individual to search for life's meaning.
When one is more reflective and appreciative of this process of
finding meaning, well-being can be enhanced, and facing difficulties
in a more mindful and meditative way (trusting that you are on the
right path) can help you overcome the health challenges.

LET'S TALK ABOUT PERSONAL DEVELOPMENT

What is your current understanding of personal development? Is it
something you've dabbled in, or is this the first time you've heard of
it? Regardless of how well versed you are on this topic; personal

development is something I encourage everyone to partake in if they desire a better quality of life. Why? Because through personal development, you get to discover and develop the best versions of yourself. It helps empower and strengthen you in ways no other education can do because it's focused on understanding who you are.

For most people, personal development refers to binge-watching Tony Robbins on YouTube or saving motivational quotes on Instagram and Pinterest. But that's barely scratching the surface of what it's meant to be. Wikipedia defines personal development as activities that develop a person's capabilities and build human capital and potential, facilitate employability, and enhances the quality of life and the realization of dreams and aspirations. The way I like to approach this is simple. Personal development is about learning how to unleash my best self. I never used to care about it until I read Dr. Wayne Dyer. I immersed myself in his teachings and programs, which slowly began my journey of growth and healing. When I read the book "*You Can Heal Your Life*" by Louise Hay, I saw the light at the end of the tunnel and finally decided it was about time I gave myself the gift of a brand-new life. Sometimes I wonder how different life would be if I hadn't combined my therapy and personal development. It's hard to be sure, but I doubt I would have lasted this long without another relapse.

The tragedy of losing my Fiancé and the despair and worthlessness I was experiencing made it hard for me to believe in a bright future. Personal development and everything else I've shared so far are what helped me piece myself back into place. So why am I sharing this with you? Because I want to suggest and encourage you to take personal development seriously as you go on this quest of healing and transformation. Steer clear of thinking about it from the superficial perspective of motivational videos and instead learn to recognize the value it can bring to the different areas in your life. Namely, Mentally, Socially, Spiritually, Emotionally, and Physically. Let's take a closer look at each area of personal development.

1.Mental personal development:

This area is focused on growing and grooming your mind. There are many ways to do this, including but not limited to reading a book, listening to audiobooks/podcasts, signing up for a masterclass, or even taking free training on YouTube.

2. Physical personal development:

Did you realize that physical activity is part of your personal growth? Real healing needs to be holistic, and that's why you need to include some aspect of physical development into your routine. Consider your eating habits, sleeping and rest, as well as body movement when you plan this aspect of your growth.

3. Emotional personal development:

Emotional development will be huge in this quest of healing because, as you realized, holding back emotions is very unhelpful. In fact, engaging in activities that help you release and process your feelings in healthy ways is, in its own way, a form of sound therapy. That's why I've encouraged journaling, thought recording, and even talking to a trusted therapist in previous chapters. But you can also track your mood or put yourself through a "mental diet" where you spend a week doing everything possible to sustain an optimistic and lively attitude.

4. Social personal development:

This area of your personal development is about enhancing your communication. Think about the areas in interpersonal skills you most need help with and come up with activities that can improve it. For example, if you struggle to listen to others, then put some effort into actively listening. If you are too shy, consider taking up a course that helps you with public speaking. What social activities can you think of that would make you a better communicator?

5. Spiritual personal development:

On the topic of spiritual development, the best thing you can do is find something that brings you peace and helps you connect to your true self. That can include activities such as reading the Bible, taking nature walks, meditating, prayer and worship, etc. I have a friend who enjoys participating in Bible Study, so twice a week, he connects with members of his church to read and discuss various Bible passages. If you're not religious at all, that's perfectly fine. As I said, being spiritual has nothing to do with religion. So, for you, it might be reading something philosophical, etc.

Here's the thing. Personal development requires a plan in order for it to yield positive results. As personal development expert Jim Rohn said, "when you look at successful people, you will almost always discover a plan behind their success. It is the foundation for success." We want you to heal and transform your life successfully, which means a personal development plan is paramount. Here are the basics of a good foundation.

Step #1: Clarify your vision

When was the last time you thought about your future and the quality of life you'd like to enjoy? Usually, when battling a mental health problem, the last thing we want is to think about the future. But that's precisely where you need to start. Choose a timeframe that makes sense for you. For some, it can be ten years out, while for others, even 12 months out is a stretch. Be thoughtful and compassionate with your current state. Understand that you can always expand on your vision as you get the hang of this. So, let's imagine what you want your life to be a year from today. Think about how you feel when you wake up in the morning. What's the first thought that passes through your mind? What's the reason you're getting out of bed? How does your body feel? How serene and clear is your mind? How is your day structured now that you changed and healed yourself? What makes you feel accomplished at the end of the day? Do you have more time to spend with friends and family? How different is your morning routine? What about your work/school? What's giving you the energy to move forward? What makes you feel successful in this new reality? What goals have you been working on that gave you great satisfaction and fulfillment?

Step #2: Become aware of your strengths and areas for improvement

Now that you've exercised your mind to identify the future experience you want to create, it's time to figure out how to map this journey. Thinking from that end result, let's figure out what you've already got going for you. Always start small, where you are, with what you have. The fact that you picked up this material to heal your current condition tells me you have a lot going for you. There's enough awareness and desire to do the right thing. It also tells me there's some untapped potential and skills you can leverage to keep you moving forward.

Think for a moment about one or two talents that you've naturally excelled at since you can remember. Perhaps there's a training you took. In my case, I realized I had lots of underdeveloped skills and untapped potential. I had given up on my psychology degree because, at the time, it didn't have as great a meaning to my life as it did after going through my devastating breakup. So, I decided to lean into that and complete my degree in psychology. Now it's your turn to pick yourself up. Write down in your journal all the skills you'd like to develop or finish developing as well as the projects you'd like to start working on to move you closer to your goals.

Next, I want you to start thinking about a handful of people you can contact to be your support structure. Don't be too specific. And by the way, it's fine if you don't have all the answers. Before moving onto the next steps, write down what you feel you're naturally good at. Can you write/sing/cook/paint/code etc.? These are all powerful allies in the making of your new life.

Step #3: Build your personalized strategy

It's time to document a simple plan that you will begin implementing daily. You have clarity on the desired future, and you understand what you've got to work with and the areas you need help with. Answer the following questions: What resources do you need to start improving areas of weakness? For example, what books can you read? Which course or mastermind can you take? Do you need to go back to school?

You also need to connect with the people who can support you. Who do you trust and respect to be a mentor? Can you hire a coach or join a community of like-minded people doing what you want to do? Write down how you plan to make this connection happen.

What timeframe are you working with to implement these new changes? I recommend creating mini milestones for the different stages of your personal development journey so you can stay encouraged. Don't forget to celebrate even the tiniest of wins, as small wins always lead to massive results. For example, when you complete a course, that's a huge win even though you're not yet at that final end result.

COMBINING THE TWO IS DEFINITELY POSSIBLE

In an ideal world, you would naturally combine spirituality and personal development for optimum results. The fact is, both of these are interconnected, and a genuine personal development quest would be incomplete without incorporating spiritual practices and awareness. Why do I say this? Because the determining factor in any endeavor you partake in will always be you. So, if you are getting in your own way, there can be no real healing or lasting transformation. That's why I encourage you to create ideal conditions for yourself that allow you to combine personal development with spiritual practices.

Why combining spirituality and personal development is beneficial:

First and foremost, spirituality involves increasing your awareness and developing a belief in something bigger than yourself. Regardless of what you call this (whether you belong to a religion or not), this level of higher understanding can be very liberating and empower you to develop a level of faith in yourself that makes you an overcomer.

Personal development and any other cognitive-based therapy you take will help you uncover all the unhealthy thoughts, emotions, and habits that have been perpetuating your mental condition. They will help you identify limiting beliefs, and as you do, that higher sense of belonging and guidance will support you as you make the shift from the old you to the new you. As the negative thoughts and limiting, beliefs are released and replaced with a higher sense of faith and positive belief, you start to experience life in a new way.

The second reason these two fields work so well together is that your ability to finally regain clarity and calmness is enhanced. The more you mindfully and spiritually work on yourself from the inside out, the clearer you become about who you are, what you want, and why you want it. That increased clarity also makes you a better decision-maker and a more optimistic individual with a new outlook now that you see life through an entirely different lens. Think of it this way. If you spent all your life trapped in a cave that only had a small window through which you could see the sky, your perception of the world would be minimal and distorted. Now imagine going from that tiny perspective (the size of a window) to having the cave walls removed and seeing the entire sky for the first time. That's what we're aiming for.

The last reason I want you to consider is that this will help you become more courageous and bolder about your dreams. You will feel worthy of setting new goals and dreaming big for yourself because you feel connected to a power greater than anything you've known. And as you continue working on yourself, having a better lifestyle will feel like the most natural desire. For example, before my breakup, there were no grand dreams or visions. The goal of marrying my fiancé was as far as I could go with my small thinking and limiting beliefs. Never did I imagine that I would end up becoming an author and impacting lives through my story, study, and research on mental health. But today, I know there's far more that I can accomplish, and my best life is still ahead of me because I keep setting bigger goals. This was only possible thanks to the personal development coaching I invested in. It took me from just wanting to be healed of my mental condition to dreaming of creating something meaningful that I can be proud of. You can accomplish the same and even more.

IMPROVING YOUR MENTAL HEALTH USING THE TWO

Here are some practical techniques you can use to improve your mental health and combining spiritual and personal development practices.

#1: Meditation

This is a great way to develop yourself and become more connected with a higher level of awareness. Research shows that meditating helps improve sleep, increases pain tolerance, reduces stress, controls anxiety, enhances self-awareness, improves focus and attention span. It promotes a healthy emotional life and can be a great way to find peace and stillness.

#2: Yoga

Regular yoga (especially yoga that incorporates deep breathing and meditation) brings about greater mental clarity and calm. It increases body awareness, relieves chronic stress, and sharpens concentration. By practicing yoga, you are getting your physical development, cognitive development, and spiritual development combined.

#3. The 555 Practice

This is a morning mindfulness practice that will combine personal development and spiritual growth, and it takes only fifteen minutes. It's a mindfulness technique created by Uma Beepat, and here's how to incorporate it. First thing in the morning, commit fifteen minutes segmented in the following way. Spend five minutes in meditation and practice deep breathing. After the alarm goes off, stretch for another five minutes and finally, in the last five minutes, mentally prepare for your day and write down your intention for the day of how you want to feel, how you will show up in the world and one or two things that you want to accomplish by the end of the day.

#4: Keep a gratitude journal

Energy flows where attention goes. That means, if you invest some time daily to reflect on the events of the day and remember the moments that made you feel good, you will generate more of those feelings and eventually stabilize that state of being. I like doing mine at the end of the day when I get into bed so I can recall even the tiniest thing that made me smile. I find that elevates my mood and enables me to sleep better. This not only develops you spiritually, but research shows it has biological and mental benefits for your brain.

10

REDUCING THE SEVERITY OF YOUR DEPRESSION & ANXIETY

Depression and anxiety can happen at the same time. In fact, many people who are depressed tend to also suffer symptoms of anxiety and vice versa. Even though these are caused by different triggers, it is common for them to overlap, making healing all that more complicated. That's why it's essential to become aware of the different ways you can reduce each of these, whether they show up singularly or together. Sometimes, one treatment isn't going to be enough, especially if you're currently in a stressful environment at work or home. So, it becomes necessary to develop tools and techniques that you can deploy in your daily life as stress triggers pop up. The goal here is to help you have as many tools and techniques as possible. Know that everything we share may not always work for every case, but it's up to you to understand your personality better so you can pick the ones that do. I like to call these coping strategies.

WHAT IS A COPING MECHANISM OR COPING STRATEGY?

It's basically a way of dealing with a problem in order to reduce stress. The more you can relieve your stress, the less severe the depression and anxiety will be. It's important to realize that not all coping mechanisms are beneficial in the long run. For example, a person may choose to use junk food as a coping mechanism, which might feel great at that moment, but in due time, that will turn into a new problem like weight issues, etc. That's why you need to be mindful of the strategies and coping mechanisms you pick. Some people go for meditation, and that's a great example of a healthy coping mechanism.

A WAY TO REDUCE ANXIETY AND DEPRESSION: COPING MECHANISMS AND STRATEGIES

When it comes to reducing anxiety and depression, you need to understand both problem-focused strategies and emotion-focused strategies. Depending on the situation at hand, you might need something to reduce stress or something to help you handle feelings of distress. If we want to effectively manage depression and anxiety, we must first realize that stress is our enemy and it is not an external issue. It's very much internal. Regardless of where you are or what's going on, if you can control your mind and stress levels, you can dampen the effects of your anxiety and depression. I know this is easier said than done, so I will share basic techniques and how to apply them. But before that, let's talk about the dark side of relying on coping mechanisms.

THE GOOD AND THE BAD

Sometimes we default to avoidance types of coping mechanisms or harmful ones that hurt us and keep us trapped in that depression and anxiety cycle. Things like smoking or using other addictive substances to ease our condition's severity only lead to more problems and are often signs of avoidance. I also see procrastination as a way of avoiding things too, and so we can also add it to this category of unhealthy coping habits. The bottom line here is that you need to exercise your power of reason and be mindful of what you choose to do when you get triggered into a downward spiral of anxiety and depression. If you want some tips of techniques to experiment with, here are several.

WHAT ARE THE 9 SIMPLE TECHNIQUES THAT I CAN DO?

The first thing you need to do is to become aware of your triggers. Once you know what triggers you, it will be easier to catch the wave before it takes you down.

So far, we have emphasized the importance of developing healthy habits and coping mechanisms. We have already mentioned things like sleep, proper nutrition, exercise, and practicing mindfulness to improve your condition. These are all highly effective strategies. I recommend incorporating them into your current plan. However, I want to take things to the next level to make sure you feel fully equipped no matter where you are or what stressful situation you find yourself in. Here are nine more techniques you can add to your self-healing kit that will improve your state and transform your current lifestyle.

#1: Deep breathing

Pause everything you're doing and give yourself some space. If you're in a room with people, try to step out to a different room or go outside and take deep, slow inhales and exhales, focusing on that breath.

Feel free to borrow the technique you learned in the guided meditation where I walked you through the rhythmic breathing practice. You can deeply and slowly inhale (one...two...three...four), filling up your lungs all the way into your belly and then hold for a count of two (one...two) then slowly exhale for a count of four (one...two...three...four). Focus on nothing but your breathing and the counting.

The first few times, you might need to do it for several minutes before you can bring back a sense of calm, but as you continue to practice this, it will get easier.

#2. Take a long nap

This might sound too simple and ineffective, but I think sleep is the best way to reset your thoughts and emotions. If you feel yourself losing control, stop everything and go to sleep for at least an hour. I also encourage you to go to bed early on the days when things feel really tough. When stressed, your body needs a lot more sleep and downtime.

#3: Practice acceptance

It's time to gain some real perspective. Instead of letting your brain and emotions get hijacked, why not get curious and question what's really going on. Is it really as bad as the voices in your head are making you believe? And even if things are bad, do you really have control over them? Sometimes that pep talk and the realization that you can't control everything is all you need to create a little more mental stability within you.

#4: Give yourself a time out

During this time out, you want to engage in something that is uplifting and naturally shifts your state. For example, if jogging and breaking a sweat usually makes you feel different, then even if you're not feeling up to it right this minute, make an effort to do a twenty-minute jog and break some sweat. You'll thank yourself later for forcing yourself to focus on something else that you enjoy instead of allowing your mind to drift and get caught up in the stress. Don't limit this to physical exercise. It can also be getting a massage, meditating, painting, coloring, poetry writing, cooking, etc. Immerse yourself for at least twenty minutes in something that takes your mind to a different level.

#5: Listen to soothing high-frequency music

If you enjoy music, it would be worthwhile to invest in spiritually uplifting music to soothe you. Research shows that the right kind of music can positively impact your brain and body chemistry. Nature sounds can also be very calming, which is why they are often incorporated into relaxation and meditation music. Have a playlist on hand wherever you go that can instantly shift your state when needed.

#6: Spend time in nature "forest bathing"

This a simple yet profoundly effective way to develop your personality and spirituality while simultaneously reducing anxiety and depression. Ever heard of forest bathing? There exists an ancient practice known as forest bath (shinrin-yoku) in Japan, which is not exercise or jogging, or hiking. It is simply being in nature and connecting with it through your senses.

Through sight, smell, touch, hearing, and tasting, you can soak yourself in nature, which bridges the gap between you and the natural world. You have to remember, our civilization is divorced from nature, and the more we advance, the harder it will become. According to a study sponsored by the Environmental Protection Agency, the average American spends 93% of his or her time indoors. This is actually detrimental to our health and well-being. That's why taking up forest bathing at least once a week is advised.

Now, I know what you're thinking. "I can't do this. I live in the city with no forest in sight." That's okay. You don't need a forest to do this. You need to find a spot (a park/garden etc.) and decide to spend an hour slowly walking and observing all the plant life and animal life around you. Make sure you have no phones, cameras, or any distracting technology. Let your body be your guide. Listen to where it wants to take you. Follow your nose and take your time. It doesn't even matter if you don't go very far. What matters is that you unlock your five senses and allow nature to enter through your ears, eyes, nose, mouth, hands, and feet. Listen to the birds singing and the breeze rustling the leaves of the trees. Observe the different greens of the leaves on the trees and how sunlight filters through the branches. Smell the fragrance of the nature around you and allow a connection and exchange of energy to form. Place your hands on the trunk of a tree. If there's a stream or river nearby, dip your fingers and toes in it. If not, find somewhere pretty and lie on the ground. Drink in the flavor of the nature around you and release your sense of joy, calmness, and oneness. You are part of this magnificent masterpiece.

There's no one-size-fits-all for doing this. Customize it to suit your needs and current situation and go where you feel most comfortable. For example, if you love the smell of water and sand, try to find a spot representing that. The effects and connection will be more powerful when you opt for environments that make you feel good.

#7: Challenge yourself to learn something new

By setting goals that challenge you, whether at work or outside (such as learning a new language or cooking a particular cuisine), you build confidence and redirect your energy. It develops your mental acuity and helps you unlock more of your potential, which aligns with both personal and spiritual development. By learning, you increase emotional resilience and arm yourself with the knowledge that makes you want to be more active instead of settling for a sedentary lifestyle. It also shifts your focus and attention from feeling sorry for yourself to empowering yourself to do something good.

#8: Help other people

Do you have a skill you can share with others online? Have you always wanted to volunteer for a particular cause? Evidence shows that people who help others through community work and volunteering become more emotionally and mentally resilient. It also takes someone's focus from their current mental problems by offering a different perspective of seeing how others are struggling (often with bigger issues). The mere act of giving your time, energy, attention, resources, and love makes you feel different and good.

I can recall a story of a member of our spiritual community who said she was having suicidal thoughts because her depression felt too unbearable but then agreed to take up just one weekend of volunteering work at a children's home. But this was no ordinary children's home. It was for kids with special needs.

Many of them were disabled and autistic. By the third and last day of her time there, she said she was a different woman. Seeing those kids who were abandoned by their parents demonstrating so much courage – the courage to try to live even when the odds were not in their favor made her question what the heck she was doing with her life. She had great parents, a wonderful upbringing, and people who loved and supported her throughout her life, even during this season of depression. Since then, she's focused on her recovery and spending as much time as possible working with disabled children. Perhaps there's a clue in there for you too.

#9: Learn to soothe and parent yourself

The full realization that you are responsible for yourself is one of the pivoting points of self-healing and mental well-being. You have everything you need within yourself and are worthy of a good life. Yes, life challenging and difficult situations will show up, but you are your own best coach and parent at this point. The most important relationship you will ever have is the one with yourself.

Make this a priority in your life and learn to speak to yourself with as much kindness and encouragement as you can. One important thing you can do at any moment to soothe yourself is to use the right words when you feel your anxiety increase or on days you don't want to get out of bed. Simple words like *"I know it's hard but just take it one hour at a time."* Or if that's too hard, you can say, *"I know this moment feels like hell but just take a deep breath in and keep breathing. You'll be okay [name]"* and call yourself by name. Always speak to yourself with compassion, love, and tenderness, especially when you're in a rough place.

Other soothing statements you can use on yourself:

- You are not alone.
- It's okay to feel this way; it makes sense to me.
- I love you no matter what.
- I'm sorry you're going through this.
- I know this is a difficult time for you.
- I'm here for you.
- You can count on me.

11

REGULATE YOUR EMOTIONS THROUGH GROUNDING

E motions are what we must understand and control because bad things tend to happen when they rule our minds. So how do we do this in a healthy way? That's what we'll cover next.

Understand that your emotions are good except when they aren't helping you. If an old traumatic experience stirs up emotions take you back into the past and cause you to fall into an unpleasant reactionary state, that has to be dealt with. A highly effective technique followed by almost all mental health professionals is grounding.

WHAT IS GROUNDING AND HOW CAN IT HELP?

Grounding techniques are a set of tools that can help us manage overwhelming emotions and traumatic experiences. Grounding is actually rooted in Dialectical Behavioral Therapy. It can be highly effective when feeling distressed, overwhelmed emotionally, or when something triggers a past memory and removes you from the present moment. I consider this a healthy way of getting back your sense of control.

What makes this work is the fact that you make a conscious effort to focus on some aspect of the physical world using your five senses instead of getting caught up in your chaotic internal thoughts and feelings. In so doing, grounding enables you to build a bridge back to this moment in time where your power lies.

By practicing these techniques, one is able to step away from negative thoughts or flashbacks. These techniques can also be used to stop the momentum of cognitive distortion before it gets out of hand. They are highly effective at decreasing the grip of unhelpful emotions. Different techniques will work better depending on your personality and emotional needs, so feel free to experiment with a handful and discard the ineffective ones. What you want is something that brings your focus back to the present moment, the current environment, detached from the past, and any negative mental movies. Let's look at

a few that come highly recommended by experts.

THE 5-4-3-2-1 TRICK

Before starting this exercise, pay attention to your breath. Take slow, deep breaths as you go through the following steps.

> **5:** Acknowledge FIVE things you see around you. It could be a lamp, a book, a spot on the ceiling, a cup, or anything in your immediate surrounding.

4: Acknowledge FOUR things you can touch in your immediate surrounding. It could be the pillow you're leaning on, the
ground under your feet, your hair, etc.

3: Acknowledge THREE things you hear right now. This can
be the sound of the air conditioner, the clock ticking, kids
laughing outside, or whatever else you can hear. Try to find
something outside of your body.

2: Acknowledge TWO things you can smell. If you're in the
office, can you smell paper? If at home, can you smell coffee
from the kitchen or your shower soap? If you can't smell
anything, feel free to stand and take a brief walk to find a scent
either in your current location or outside.

1: Acknowledge ONE thing you can taste. If you like, you can
pay attention to your saliva or the inside of your mouth, especially if you recently drank or ate something.

PLAY THE MEMORY GAME

Activating a positive memory can help you escape unwanted negative
emotions. To use this as a grounding technique, consider looking at a
detailed photograph or picture of scenery that stirs up positive
emotions. Do this for five to ten seconds. Then turn the picture face-
down and recreate that same image in your mind in as much detail as
possible.

Another way to approach this could be to find a picture with lots of objects, e.g., a busy city, the interior of a beautiful home, etc., and stare at it for ten seconds, then turn face-down and mentally list all the things you remember from that picture. I have been using this for years, and it works every time. I carry lots of inspiring photographs on my smartphone, and whenever I feel myself drift, I find a photo of somewhere exotic that makes me feel good, and then I mentally list everything that's on the picture. It takes less than a minute, and I am back in the present moment.

OTHER GROUNDING TECHNIQUES TO LOOK INTO

#1: Progressive Muscle Relaxation

This technique involves you relaxing all the muscles in your body. You can get a guided progressive muscle relaxation video on YouTube to walk you through it or simply borrow my tips. First, Tense and relax each muscle group, head to toes or vice versa. Do it, one muscle group at a time. Tense for five seconds, then let go and relax all the way, paying particular attention to the difference between tension and relaxation. As you relax, keep repeating the simple statement "relax."

#2: Recite something

You can create your own mantra or memorize a poem that you care deeply about. If you're religious, consider learning a passage from the book of Psalms that is powerful enough to refocus your attention and energy. Recite this poem, song, mantra, or Bible verse to yourself or mentally in your head. If you do it mentally, please visualize each word as you'd see it on a page. If you choose to speak the words aloud, focus on each word's shape on your lips and in your mouth.

#3: Visualize your favorite place

If you find it easy to connect with places visually, this can be a cool technique to use in seconds. Think of your favorite place. Somewhere that always makes you feel safe, loved, and at peace. It can be your childhood home, current residence, a vacation in a foreign country, or anywhere else you've been where beautiful memories were created. Use all your senses to recreate that mental image. Think of the colors, sounds, sensations, and how it felt to be there. Who were you with, if anyone? What did you do there? What made you most happy during that time?

#4: Soothe yourself through your senses

Your five senses can be a great way to bring yourself back into a calm, present state. Here are the different activities for each of your senses that can act as a soothing strategy while simultaneously promoting self-care habits.

Sense of touch can be soothed by getting a massage, soaking in a warm bath, going for a swim, soaking your feet, wrapping yourself in a cozy blanket, putting a cold compressor on your forehead, cuddling with a lover, hugging someone you like, putting on lotion on your body, playing with an animal, and stretching.

Sense of taste can be soothed by sipping your favorite herbal tea, chewing a piece of gum, eating your favorite nutritious meal mindfully, slowly sucking on hard candy.

Sense of smell can be soothed through deep "belly" breathing, lighting a scented candle, using an essential oil diffuser, shopping for your favorite flowers, using your favorite soap, shampoo, lotion, or perfume.

Sense of sight can be soothed through reading a book, watching the clouds, lighting a candle and watching the flame, a beautiful flower, watching the sunrise/sunset, looking at pictures of a loved one, past vacation, or a place you dream of visiting.

Sense of sound can be soothed through listening to uplifting music, humming or singing to yourself, playing a musical instrument, getting still and listening to the sounds of nature around you, saying positive statements to yourself, or self-encouragement.

Now I invite you to make a list that will form part of your distress kit to help you when your emotions try to kidnap your brain.

WILL ALL OF THIS WORK?

Being able to ground yourself is just as important now as when you were a child. The only difference is that you had to depend on someone else as a baby, whereas now, you need to take full responsibility for this. The more you learn to ground yourself and maintain calm in the face of chaos or a disturbing emotional experience, the more resilient and mentally healthy you will become now and in the future. This isn't just about healing your condition now. It's about giving you the tools you can continue to use for the rest of yourself to avoid any relapse.

I should remind you that not all grounding techniques will work. Different personalities will find some methods work better than others. What I've shared here is only a handful. So, if you experiment with these long enough to no avail, keep looking for and testing new ones until you find the right fit.

WHAT TO DO IF NONE OF THESE TECHNIQUES WORK:

The worst thing you can do is panic, get frustrated, or fall into negative self-talk, thinking something is wrong. There's a lot of information on the Internet with dozens of other shared ideas. Keep researching, and also check out my resource page at the back of this book for more guidance on where to find other grounding techniques.

Grounding yourself will always work, and it's a great approach to managing your emotions long-term. You just need to find the best techniques and practice them long enough.

III

WHAT NOW: TAKING THE
NEXT STEPS TO RECOVER &
HEAL

GOAL SETTING: MAKE IT PRACTICAL AND MEANINGFUL

Let me ask you this. How often do you begin any endeavor with a clear goal and end result in mind? I don't know if you noticed, but this book is intentionally designed to make you more vision-focused (on that ultimate result of a healthy, happy, successful life), none of which can be possible without proper goal setting.

Goal setting is touted as an essential step to becoming successful in the personal development space. But I want to show why it's also vital in your healing process whether you choose to self-heal, hire a therapist, or combine both approaches.

WHY IS GOAL SETTING SO IMPORTANT?

Think of it like this. You live in Florida and desire to drive to New York for the first time in your life. You have the end result in mind, but without a clear map of how you will get there, you might end up on the opposite side of America, frustrated and feeling like a failure. The same can be said of your quest for a new lifestyle free from the torture of mental health problems. You now have a vision of what you want your life to be like a year, two, or even three years from now.

That's your destination. What's needed now is a strong strategic plan of how to get there. This plan will include little touchpoints or milestones that can help you track progress and validate that you are on the way to that grand vision. These are the goals we want to identify and bring them to the forefront. So, when was the last time you took some time to map out your health goals or any other goal for that matter?

Why Do I Need to Set Goals?

Research shows that setting a specific goal makes us more likely to achieve the things we want. It is imperative to set clear goals when we are seeking to make significant shifts in our lives. I consider health goals the best starting point for you because it's a way for you to use this tool to turn something negative into a positive while at the same time proving to yourself how powerful you can be. By setting a goal and achieving it, you literally show yourself that you are capable of achievement. And as you keep piling on wins and hitting one milestone after the next, your self-esteem, confidence, drive, and enthusiasm increase naturally. If you want to set your life on fire and live on a good high, follow the ideas shared here, start small, take small, consistent steps toward your goal without getting distracted or discouraged, and you will permanently transform your life.

SHORT-TERM AND LONG-TERM GOALS: ARE THEY NECESSARY?

If you ask a large enough audience about goal setting, you'll come out confused. Some will swear against goal setting, while others would never go through life without clear goals. I think the reason for this divide is that many people go about goal setting superficially and half-heartedly. When you create any kind of plan, you need to commit and stay focused until you've attained it. You also need to reach for goals that are meaningful, aspirational, transformation, and measurable in some way. In that sense, setting a goal to get rich for the sake of getting rich wouldn't work out well because it fails on most of the criteria I just mentioned. Healing yourself from major depression so you can become a stable and better parent, spouse, child, or boss makes a great case and a powerful goal.

That's why I suggest you identify your big goal, which is the more long-term plan that will bring your vision into reality, as well as short-term goals that move you toward the attainment of that grand goal. For example, you have battled depression all your life and now struggle with the negative impact it has on your new relationship and current job. Despite all your attempts at containing it, things keep getting out of hand, and you need a new life. You're fed up with being a slave to emotions, the economy, etc., and always at the mercy of the next therapist, so your grand goal is to be 100% healed from depression and the tendency to relapse. That's a noble goal for yourself, your family, and those in your circle of influence. But how do you actually get to that goal? By identifying all the mini short-term goals that can enable you to move toward and eventually attain that big goal of being a healthy, successful husband so you can finally start a family of your own. Short-term goals would include like, read books on self-healing (which you're already doing, so congratulations on taking that first step), start exercising five times a week for half an hour, reduce my caffeine intake to one cup of coffee in the morning, stop drinking beers every night, etc. I think you're catching the gist of how this works. So, let's outline some strategies you can apply as you map out your short and long-term goals.

STRATEGIES FOR SETTING GOALS

Let's reflect on a few critical questions: In what ways do you want to improve your mental, emotional and physical health? Which bad habits are you ready to change? What do you wish to improve about your relationships? What skills do you want to learn? Are there any other things you've been thinking about changing in your professional life, social life, or financially that you believe will help you manifest your ideal lifestyle?

Once you have a few things in mind, pick the highly charged one that feels most pressing and directly aligned with that vision of your new life and use the questions below to fine-tune whatever came to mind. I recommend journaling this down somewhere private and easy to access.

#1: Write down all the hot ideas that came to mind as you read through the reflective questions above. Don't worry about ordering them in importance, just put them down on paper.

#2: Take what you wrote and order them in importance, looking for the one that lights a fire within and makes you want to take action now.

#3: Check to see whether you have written these in a way that makes them feel real and achievable for you. Also, make sure the statements are positive meaning, it should be what you want more of, not what you don't want. For example, if you're going to heal from anxiety disorders, instead of stating, "I want to be less anxious," you could say, "I want to be more relaxed, calm, and "clear-minded. See what I mean?

By the way, it's okay if you struggle to find what you want when starting this exercise. If you feel like you don't know what you want to set as a goal, start by doing a simple exercise called clarity through contrast. In this process, you get a piece of A4 paper and split it down the middle. On one side, label it "What I don't want," and on the other, "What I want." Begin by listing down everything you don't want. Once done, shift to the other side of the paper and find the antonym or opposite of what you wrote down. Word it in a way that makes you feel good. Now transfer the new wanted desires to your journal and proceed with step #3. Shred or burn the A4 paper. You no longer need to worry about what you don't want.

#4: Refine your desires and goals in more specific ways. For example, if you have as your top goals "to be happier," try to challenge that statement until you get something more specific. What does happiness feel and look like for you, and what actions or behavior would indicate that you are indeed becoming a happier person?

Once you are done with this, it's time to take massive action on your goals. We do that through a detailed action plan. This includes writing how you're going to measure progress, the timeframe you're giving yourself, and any other essential details. For example, if you realize your current goal needs to be broken down further into even tinier steps, this is where you map that out. I suggest goals to be 90 days or less so you can easily track progress and feel like you're moving in the direction of your vision as you review every ninety days or so. That kind of a goal might still require mini milestones broken down into weekly micro-goals. That's a great way to keep yourself on track, too. Just be sure to specify what you expect from yourself within a week and try not to make it too huge a hurdle. Now that you've broken the goals down into micro-goals, what are the next logical steps you can take today? Remember, it should be baby steps. Think of three things you could do right now to move you toward that 90-day goal, which will ultimately lead to your new lifestyle a year from now. A few more things to consider is having a vision statement written out somewhere visible so you can see it daily. I also suggest you write out your goals each morning so you can spend time immersed in this new momentum that's building at the start of your day. Last but not least, please start building a small list of the resources and people you would like to have as your support along the way.

ACHIEVE YOUR GOAL BY BEING SMART

Before moving on, I wanted to share another popular method for goal setting that you can use if you prefer a more structured approach. It's called the SMART technique. In Cognitive Behavioral therapy,

SMART (specific measurable achievable realistic and time-framed) goals are often promoted. Of course, this method may not work for everyone, but if it resonates, here's how to do it right.

S: Specific means you need to set a specific goal for yourself. If you've hired a therapist, they can help guide you through this. In essence, you want to make sure the goal is as specific as it can possibly be.

M: Measurable means you get to track progress either daily, weekly, or monthly so you can know if you're on the right and what's working. The more specific you are with your goals, the easier it is to track and measure them.

A: Achievable simply means it needs to feel like you are capable of doing it. And this can be tricky because oftentimes, battling mental health problems for long periods leaves us feeling defeated and incapable of doing anything praiseworthy. We usually lower our standards too much, leading to a cycle of low self-esteem, which is poisonous during this process.
So, while I will caution you against wishful thinking, I also want to challenge you to make your goals big enough that they excite you. You'll know it's right when you feel a little excited, and a bit scared at the same time.

R: Realistic means choosing a goal that is possible. For example, going from major depression to 100% in a week is so unrealistic. Be mindful of the timeframe, the plausibility, and your ability to commit to the end as you settle on the details of the goal.

T: Time frame involves establishing a precise amount of time that you will dedicate to this process of achieving your goal. Most people usually underestimate how much time they need to reach their destination, leading to discouragement and failure. As you decide how much time you will give yourself, remember this should be a guiding line, not a determinant. If you set 90 days for goal completion, and after that duration, you're still at it, then so be it. Carry on with no less enthusiasm and determination than when you first started. If you've done everything else outlined in this goal-setting chapter, success is inevitable as long as you don't stop along the way.

13

WHY TRACKING & SELF-EVALUATIONS ARE IMPORTANT ON YOUR HEALING JOURNEY

You've learned about SMART goal setting and the importance of measuring your progress regardless of the goal in mind. I want to emphasize this idea of tracking and measuring progress specific to your healing and recovery. As we know, things often feel stagnant, which causes many people to quit their treatment. One of the main reasons people quit their treatment is that they think it isn't improving their condition, whether it's anxiety or depression. This is a feeling you can avoid if you start to track and self-evaluate.

TRACKING AND SELF-EVALUATION, YOU NEED IT

Self-evaluation is the process of systematically observing, analyzing, and determining the value of your actions and consequent results in order to stabilize and improve it. While it can be applied toward many different aspects of your life, using this procedure in your therapy can help you manage it better and stay on track as you witness progress. It also helps to keep you accountable so you can continue to work on yourself and complete assigned homework if undergoing CBT, DBT, or ACT with a health professional. The primary purpose and why you need this is to highlight the wins, accomplishments, and good work you're doing. You have to be proud of what you're achieving week by week as you build toward your new lifestyle. Tracking and self-evaluation are beneficial to people who have a tendency of rational or analytical thinking as opposed to those that are the quantifying type. So, knowing more about your personality does help a lot. By tracking your treatment, both you and your therapist can have a baseline that can be used when things get off track.

How Can it Help Me in My Healing Journey?

Therapy is often like trying to sort out and complete a complex puzzle that you're carrying. There are many approaches that you and your therapist can apply depending on what feels right for both of you. Once you identify the final outcome, how do you get there? And better still, how do you show evidence that you're getting closer to that chosen destination? With a tracking tool, you are able to both establish where you are and where you want to go. Having a symptom analysis system could help prevent you and your therapist from feeling stuck, and it could substantially reduce ambiguity. Most therapists will have their own checklist, assessment tool, or even a mobile app that you could use, but even if you're not working with a therapist, you can develop a method of tracking progress.

THE DIFFERENT WAYS TO TRACK MY JOURNEY

The most standard approach is a written service plan with goals and objectives that you've identified. Your therapist may determine progress based on the achievement of goals with quarterly updates to those goals. You could also combine the treatment plan with rating scales and other short, standardized assessments to track symptoms over time. Certain therapists are more technologically advanced, whereby they have mobile apps that track symptoms daily and produce reports for you as well as some standardized assessments that help you know how you're doing.

Suppose you also feel the need to track habits that promote healing, such as exercise, meditation, sleep, and so on. In that case, you could additionally use mobile apps that give you daily, weekly, and monthly reports on your progress. There are other ways to keep tabs on your progress, including:

Personal journaling: For this to work, you need to do it long term and create a consistent routine. For example, you could spare five minutes each morning to write down how you're feeling. If it's the start of the day, write out how you hope the day will go and if it's at the end, write what worked and what you can feel good about. Do this for at least 12 months, and you'll have powerful resources that help you see how far you've come, mostly on the days you feel extra vulnerable.

12-hour check-ins: Think of whichever mode of communication you enjoy (writing, voice, video, etc.) and commit to recording something short every 12 hours to update yourself on how you feel. If you find that you're continually waking up in the middle of the night anxious, record that. This will enable you to become aware of triggers, patterns, and things in your life that are either promoting or hindering your recovery.

SELF-EVALUATION METHODS THAT AVOIDS BIASES

If you're going to create your own tools for self-evaluating and tracking progress, make sure there is no room for bias. Don't shortchange your recovery by cutting yourself some slack. The tools and methods you use should be easy to follow, balanced, and concise. So, aim for something straightforward, fair to your current situation, and meaningful. You could even start by testing pre-existing large-scale assessment tools and practices and use the best ones to inspire your own. The self-assessment should empower you, not deflate your self-esteem, so whether you establish your own or work with a standardized one from a therapist, use tools that remain objective and progress oriented.

14

INTO THERAPY: SHOULD I GET IT?

Although this book is centered on a self-healing approach, we must recognize that the psychotherapy discussed, i.e., CBT, DBT, and ACT, traditionally require a therapist. Reading books like mine and doing everything you can to practice self-care should not be diminished in importance. Many have self-healed from phobias, anxiety, and eating disorders without seeing a therapist. Still, for the most part, if you feel your condition is pretty severe, you're going to need a combination of processes to get you better.

Therefore, we must at least discuss the benefits of hiring a therapist and partaking in a more conventional treatment regimen as you learn to self-heal. Signing up for a program with a therapist doesn't contradict what is shared in this book, and it certainly doesn't make you any less responsible for your healing. It offers you a safe space and a trusted partner who you can work with and keep you accountable during this process. But a lot of people still struggle with the thought, "do I really need therapy?" After all, how many times have you heard someone bashing on therapists and how horrible therapy was? So, does it really work?

THERAPY: WHAT GOOD DOES IT DO?

While I am not convinced that everyone suffering from mental health problems needs a therapist, I believe most do. Most of us could use that little extra support no matter how great our coping strategies are or how big our distress kit might be.

Dealing with our own thoughts, feelings, and behavior isn't always easy so before you dismiss this idea, hear me out. Research has shown that verbalizing feelings can have a significant therapeutic effect on your brain. Some might think this makes you weak or that you'll come across as a wuss, but that couldn't be further from the truth.

I recently read an article of a guy who shared that he ghosted his therapist and a few weeks later started regretting the decision because things actually got worse. His main conflict was that he thought it made him look weak to his friends (as though there was something wrong with him), and that conflict made it hard for him to honor the agreement with his therapist. Unfortunately, he realized only after the fact that he was better off having that kind of a support structure in his life. It's quite sad that our society still perceives therapy as something negative, and hopefully, your friends are smart enough to see that you seeking help actually proves that you are healthy.

I have grown to see psychotherapy as a valuable tool that can help us become successful, healthy, and happy individuals. Every approach you've learned in this book can be broadly categorized as "talk therapy," which I feel is the best when combined with self-healing. The trick, however, is to find someone you connect with. Someone you think you can trust. There are many benefits of opting for professional help, including but not limited to:

- You get a whole new perspective on yourself and who you are.
- It gives you a different understanding of others and your environment.
- By sharing what's going on inside in a safe, judgment-free environment, you start releasing those strong negative emotions.
- It helps you deal with future curveballs and challenges that life will throw at you.
- You get validation that you're not crazy or broken and that you can fix whatever issues have been ailing you.
- Any physical trauma and other related physical ailments get treated as well.
- The positive effects of talking with your therapist gets internalized so that self-therapy picks up where the actual therapy leaves off.
- Being in therapy helps you improve your communication skills.
- You feel empowered and supported to make healthy choices.
- Therapy makes you feel empowered.

WHEN SHOULD I GET THERAPY?

The simple answer is that you ought to consider seeking professional help if you're struggling with emotional difficulties, if you're unable to function normally in your daily duties or when you start having mental health concerns. I don't think anyone can really tell you that you "need" therapy, but if you are experiencing the following signs, I encourage you to reach out to a professional.

#1. Your performance is significantly down whether at work or at school.

#2. Your sleep is drastically disrupted. Either you're suffering from sleeplessness or fatigue and just want to stay in bed all day.

#3. Your appetite and eating habits feel out of control. You might be stress eating and using food to dull your emotions, or it could go to the other extreme where you barely eat and end up starving your body.

#4. You've become a recluse and have difficulty maintaining any meaningful relationship. You might find yourself often in conflict with others or unable to effectively communicate.

#5. Things you used to enjoy now feel like a big burden. There's an emptiness inside that won't go away.

#6. You're becoming dependent or even addicted to alcohol or other substance or maybe even using sex as a coping mechanism.

#7. Disproportionate anger, rage, or resentment. Especially the ones that reoccur and get out of hand even with trivial things.

#8. Intrusive thoughts that seem to hijack your mind frequently.

#9. Physical health issues. If your health has taken a considerable dip, it could be a sign of underlying mental issues.

#10. Hopelessness and maybe even suicidal thoughts should be a huge alarm that you need professional help.

WHAT MAKES A THERAPIST GREAT: CHOOSING THE RIGHT ONE

A tremendous amount of research proves that healing in a patient is influenced by the therapist-patient relationship. That means choosing your therapist shouldn't be rushed or superficially done. You must find the right therapist for you. Just because a friend tells you to use their therapist doesn't mean it will work out. So how do you find a good therapist?

The first place I would start is by asking people you know and trust. This gives you an easy starting point and lots of options to conduct a thorough interview.

If you have a family doctor, that would also be a good source. If you have none of these options available, then my second suggestion would be to go online.

Conduct thorough research to find top recommended psychotherapists in your area specializing in what you feel you need help with.

For example, if after the self-diagnosis we conducted earlier in this book you realize you're suffering from major depression, you might want to look up the best mental health professionals that specialize in that. There are also websites of national mental health professional organizations like the American Psychological Association that can help you find a therapist near you.

What makes a great therapist?

Depending on your personality and preference, your criteria may differ from mine so go with your values and what you care about. But here are some fundamentals that I think apply across the board.

1. Seek out a therapist with the right qualifications and someone who is experienced with your specific problem. Therapists often attach a lot of ambiguous initials to their names, which can be very confusing. Here are some terms to familiarize yourself with.

- PsyD stands for Doctor of Psychology.
- LMHC stands for Licensed Mental Health Counselor.
- LPC stands for Licensed Professional Counselor.
- NCC stands for National Certified Counselor.
- LCSW stands for Licensed Clinical Social Worker.
- LMFT stands for Licensed Marriage and Family Therapist.
- LCDC stands for Licensed Chemical Dependency Counselor.
- MD stands for Doctor of Medicine.

2. Choose a therapist who has been practicing in the field long enough. I think ten years is ideal.

3. Make sure the therapist has a good reputation and follows guidelines and a code of ethics. They should be licensed (registered) in the state or territory in which he or she practices. You can research to make sure they've passed their licensing test and background check.

4. A good therapist will also maintain continued education credits and stay on top of the latest research regarding the specific problem you need help with. So, ask for a professional bio if it helps you learn more about the person and their experience in the field.

5. Seek to hire someone who shows warmth, genuineness, and empathy. Typically, a therapist is willing to a brief consultation over the phone as an "interview." This is an excellent opportunity to verify their character and to see whether you're a good match.

You should feel free to interview as many therapists as you like before deciding whom to work with. The best fit should be someone you feel comfortable with, so note what you prefer in terms of gender, age group, religion, or values. There's no need to hire someone with fifty years of experience if you can't stand talking to older people and certainly no need to interview male therapists if your religion doesn't allow you to speak openly with members of the opposite sex. So, what I'm saying is be sensible about your choice and trust your gut. During the interview, you'll want to make the most of that time so you can leave feeling informed and ready to make a decision. You can adapt the questions below to ask your therapist.

- How long have you been practicing?
- Have you seen a lot of clients with similar concerns to my own?

- When was the last time you treated someone with a problem similar to mine?
- My problem is [insert yours]. How would you go about treating that?
- Do you tend to lead the session or follow my lead?
- What are your strengths as a therapist?

WHAT TO EXPECT DURING A THERAPY SESSION

The first appointment is usually nothing like follow up sessions because this is the first "intake" session you'll have where you become formally introduced and work out the logistics of this new arrangement. Your therapist will likely explain how therapy works and provide some forms for you to fill in. You'll discuss confidentiality and how the rest of the sessions will go. If you've already got information regarding payments, they will be concluded here in this first session, or it might be when you're advised on the financial aspect and how to proceed with payments.

Once the logistics are out of the way, you'll be encouraged to share your problems, the symptoms you're experiencing, the goals you have in mind, etc. That will open up the conversation around your medical history, childhood, family history, or any mental health treatment of the past. Having this information will give the therapist an overview and help them understand where to begin. If the meeting is taking place online, this process may vary, so understand this is just the standardized approach. Another factor to bear in mind is that how you foot the bill will influence this first meeting more often than not. For example, people paying through an insurance company would first go through an assessment conducted by the therapist to identify whether they meet the criteria for a mental health diagnosis (e.g., depression). So, if you're paying cash, you're likely to have a different experience. Use this first session to ask as many questions as you need to establish that rapport and to feel confident that you've made the right choice. Get clear on the nature of your relationship and establish a baseline of how things will work between you and the frequency of your meetings, and how you'll track and measure progress.

How Do I Know if It Works?

Even after going through the daunting process of interviewing and seeking out the best fit for you, that doesn't mean you will see a difference overnight. Therapy requires commitment, time, and patience on your end. Half the time, it won't be evident to you that something is improving.

I can recall when a friend of mine said, "I like my therapist, but I have no idea what he means when he says I'm making progress because, to me, it still feels like I am right where I was three months ago." And so, I asked him to walk through his private journal where he had been tracking his habits each day. We found he started the process by recording each morning, "I hate my life. I don't want to get out of bed. I wish I didn't have to go to work." Gradually, those morning sentences started shifting. The most recent morning log actually said, "I'm feeling really vulnerable today, but I think I'll be fine. Looking forward to my hour of tennis with Jim this afternoon. I will crush him today." It may not be the ideal life my friend desires, but given where he was 12 weeks ago, I'd say his therapist is spot on.

Therapy doesn't work like medication or drugs. There's no instant gratification. It's a process, which's a good thing because it resolves deep issues and permanently transforms the individual if done the right way.

So, if you're struggling with this or wondering how you'll know that it works, I encourage you to hold the long view and do what I've recommended several times – track and measure. Use the SMART goal setting method and let your therapist to help you establish the big and micro-milestones that can help you stay encouraged and know that it is working. Although therapy is a subjective process, you can still find ways of measuring your progress. If you have a great therapist, they should be able to create milestones and goals that keep evolving and adjusting as you go through treatment.

Ultimately, measuring how successful your therapy is comes down to noticing the difference in your thoughts, feelings, and behavior. Are your symptoms getting better? Do you manage your emotions a bit better than before? Are you feeling more aware and able to direct your attention and emotions?

If you went to therapy because you were dealing with anxiety issues, have they decreased in intensity? Are you able to function with your day-to-day activities without being hijacked by panic attacks? Do you have more nights where you can sleep peacefully without those middle of night panic attacks? Are you sleeping better?

You could also use a tool like the Patient Health Questionnaire (PHQ9), enabling you to monitor your symptoms. If you've got your own app or enjoy journaling like my friend, that could be a great way to track the progress. Keep in mind that it won't be a linear experience. Some weeks will still be hard. The dips or plateaus you might witness as you track do not imply that it's not working. That's why I insist on keeping the long view. Use the vision exercise we went through and keep your eyes on that one, two, or three-year vision. Be less concerned about the day-to-day temporary dips and more obsessed with that long-term vision. Above all else, learn to trust that it will work because you've decided it must work.

15

GOING FORWARDS SOMETIMES MEANS TAKING A FEW STEPS BACK

R ecovery and permanent healing, or at least a fully active lifestyle, is what we all desire. The journey to that state will come with some obstacles that you need to prepare for to avoid premature abandonment of your treatment. This last section will prepare you for the aspects that few are willing to talk about – the setbacks.

Whether you're at the early stages of your treatment or already completed your treatment and enjoying the new state of sound mental health, setbacks should not wear you down. Yes, it's discouraging when they occur, especially if you thought you were finally past that tendency to relapse. But here's the thing. We all experience some form of setback, whether minor or major, at some point in our lives. And even people without mental disorders suffer setbacks when going after their dream lifestyle. I hope you will find the information in this chapter especially encouraging so you can have it as your reminder the next time you catch yourself feeling down or about to fall into that pit of despair.

The best place to start when it comes to preparing for a setback is by understanding what it is and why it might show up.

SETBACKS AND RELAPSES: WHAT ABOUT THEM?

A setback can be defined as a moment or period when you seem to deviate or "go backward," hindering your desired outcome of being fully healthy and healed. For example, if you suffer from anxiety disorders and go a year without any panic attack, then one day a situation triggers your anxiety, and you go into a massive panic attack, that might be considered a setback.

As many of us know, a relapse is that recurring cycle of getting better and then falling back into the same problem (in my case, it was depression). With mental illness, when the condition is managed, things can feel really good, and you can feel strong and normal again, only to relapse as soon as a major event puts you under high-stress conditions. In my experience, each time I had a relapse proved worse than the previous episode. Real healing cannot just be focused on recovering from your condition; it should also focus on preventing relapse.

Why Do We Have Them?

Sometimes setbacks and relapses can occur because you change your medication or suddenly stop taking a certain drug. That's why you should always consult your psychiatrist before making any changes to prescriptions. In other instances, it can be caused by an external trigger that stresses you out beyond your ability to manage your thoughts and emotions. For example, Lucy shared her story with our online community of how she was discouraged by her setback not too long ago. Lucy had been struggling with severe anxiety disorders since she was thirteen. Her parents had taken her to several psychiatrists, and over time, it seemed to be under control. One day as she drove on the highway, she felt dizzy, and since she was currently sick with a chronic sinus infection, she immediately started panicking. She thought she would pass out, and her car would career off the road. Somehow, she was able to get off the highway, slow down on the side of the road a few miles from home and called her mom to come to take her to the hospital. After that, it became impossible for Lucy to drive again. Each time she tried; thoughts would haunt her. She would watch the rearview mirror and imagine cars crashing into her, which got her really scared and dizzy. Feelings of disconnect would overwhelm her and eventually turn into a full-blown panic attack. It wasn't too long before she was avoiding all the highways. Of course, she realized this wasn't sustainable, especially if she wanted to keep her new job, so she sought help. She wanted to overcome her anxiety once and for all, but more importantly, she wanted to stop living with the fear of getting another episode.

Most of us can relate to Lucy's story, but the key thing to takeaway is that having a setback like her doesn't mean there's something wrong with you. The only thing you must do is liberate yourself from that fear of getting a relapse because that keeps you hostage and hinders full recovery.

What do I mean?

There are many ways to approach this. You could entertain the negative thoughts that tell you this isn't working or that you messed up. But that won't be helpful or healthy. Understand that recovery does entail having days where you fall back into irrational thoughts, negative emotions, and even habits that you know are no longer good for you. For example, if during your treatment you have a streak of days where you wake up and don't feel like life is hopeless and you actually manage to get out of bed and report to work on time, then yes, you are making progress. But suppose a few weeks in, you wake up, and your depression immediately takes over, making it impossible to get out of bed. Does that make you a failure? Absolutely not!

You are not stuck in that state. That's just the old state that is still trying to fight for your attention. Instead of assuming you've failed, and the treatment will never work, how about soothing yourself into a better feeling state? Create some bridge thoughts that can help you still face the day. For example, you could say to yourself, *"Today is going to be an off day for me, but it's okay. I know what a good day feels like, and it's okay that today I can just take things slow and pace myself. I will take it one hour at a time and just focus on getting out of bed and making myself some herbal tea. I know I'm not at my best today, but at least I am trying. I've got this."*

CAN I PREVENT IT?

Old habits die hard. There's not going to be an easy way of dealing with a relapse once it happens. But here's the thing. It is unlikely that you will experience a relapse anytime soon, so you need to release that fear at least for the next year or two. Why do I say this?

The healing process doesn't happen overnight. Even with therapy and self-care strategies for healing, you must give it time. Although no one can tell you exactly how long your recovery will take, I can assure you whatever happens in the next six to twelve months will still be part of the healing journey. So, if you stick to your new habits without fail and create a six-month streak of no episode or major issues, then one day you wake up, and it's a nightmarish day where everything seems to go wrong, that shouldn't discourage you. It's not really a relapse because it's still part of your healing.

There will be days where you will take three steps forward and days where you will take two steps back. All this is normal, and you must embrace this process. Be compassionate with yourself. Give yourself ample time to permanently heal. The more you set the right expectations, the easier it will be to avoid misinterpreting the situation and making things worse.

If, however, you really do find yourself right in the smack of a relapse, I want you to be kind and loving. Understand that what matters isn't that you never relapse or fall ill again but that you feel confident enough to pick yourself up and keep fighting the good fight. Mental health problems are like massive wars that we are fighting. They train us to become warriors. And as a warrior fighting a colossal war, some battles may knock you down. When you take a hit, you take a hit, but you never take your eye off the end objective. You know, losing one battle in the grand scheme of things is manageable as long as you never surrender. Make peace with that situation, and as quickly as you can, start getting back up!

I have trained my brain to view relapses as a hidden opportunity for further self-exploration and personal growth. Instead of perceiving myself as a failure when I catch myself falling off track, I force myself to get curious about what's happening. And on the days where it's too late for me to prevent some of my old behaviors from taking over, I usually find time afterward when I start to feel a little better to ask myself empowering questions. Remember when I shared that I used to ask myself, "why is this happening to me?" Nowadays, that irrational thought doesn't even pop up during a moment of distress. Instead, I ask questions such as "what have I learned from this?" and if it was a particular trigger that I missed, I will be asking myself, "how would I like to deal with that trigger next time?" This didn't happen automatically. The small baby steps that I have taken over the years, the same ones I encourage you to take, led to this new lifestyle that I get to enjoy, and the same will be for you. To help you develop a useful framework that can alleviate the fear of a relapse, here are the most common triggers you need to become aware of, plus tips on how to avoid each.

STRESS

It's no secret, stress tops the list. Research indicates that when a stressful situation arises, people suffering from any kind of addiction tend to crave their addictive substance or activity the most. Even if you're not an addict, stress will likely trigger your mental disorder if left unchecked.

What to do: The best solution is to avoid situations that are extremely stressful until you learn how to handle your emotions. It might help to make a list of the places, people, and things that cause excessive stress in your life. For example, if you're in a toxic relationship, it might be a good idea to distance yourself or end that relationship all together so you can focus on your healing. This is part of making that all-important lifestyle change necessary for a full recovery and a better quality of life.

In some cases, you may not be able to just eliminate the extremely stressful situation, e.g., if you work for a boss that triggers you in all the wrong ways, I wouldn't necessarily suggest you quit your job because being unemployed and losing your health benefits and salary might trigger even more stress. Instead, I recommend training yourself to manage stress better. Many personal development teachers share helpful natural ways to manage stress, including Dr. Deepak Chopra. If you resonate with him, I suggest subscribing to The Chopra Well blog and YouTube channel so you can get all the free resources and meditation practices for stress management. You can also practice mindfulness and the various mindfulness techniques you've learned in this book when you are in that stressful place.

NEGATIVE EMOTIONS AND IRRATIONAL THOUGHTS

Unchecked emotions and irrational thoughts can trigger your relapse if you don't act as soon as you become aware of them. Here's the thing. No matter how much therapy and mindfulness you practice, you cannot completely eliminate negative emotions from your life.

None of us can. Negative and challenging emotions that make us uncomfortable are part of everyone's life. There will always be sadness and darkness, and negativity in the world. But that doesn't mean we are at the mercy of every negative emotion.

What to do: Learn to process your emotions and to see feelings for what they really are. Even those that are dark and often lead you into despair do not have power over you. If you can become more comfortable with the uncomfortable feelings and approach them more as an observer than a victim, they will flow in and out of your experience quite naturally. And suppose you stay calm during that temporary period of emotional disturbance instead of going into panic mode, assuming it to be a sign of impending doom. In that case, you can release them pretty quickly.

When they come, and they will - try journaling, meditating, praying, or even sitting in total silence observing the chatter and disturbance in your mind. Find a healthy way to release the negative state.

LONELINESS AND LACK OF SUPPORT

Few people talk about this aspect, but it can be a massive cause for relapse. Most of the time, the old habits and old friends will become an obstacle to our recovery. Many experts advise letting go of old friends that only help anchor in the condition and instead work on finding a new positive support group.

In the case of depression, most of us barely have real friends, so the main issue is feeling supported by people who genuinely understand you. Regardless of where you stand on that spectrum of friends (whether you have toxic or no friends at all), you need to feel supported and loved during your recovery.

What to do: If you have made the wise choice of getting therapy, make sure you stick to it until completion. Through your therapist, you can build a healthy support system, and that will create a safe space even outside the therapy room. If you're only opting for the self-healing approach, then search online or in your local area a group of like-minded people that you resonate with. If you feel trust and genuine care within your group, then you're in the right place. The people you want are friends who can uplift you and keep you accountable when you need it. Trust me, on the journey of healing, you will need people you can trust around you.

GETTING BACK ON TRACK

No matter how great and detailed your SMART goals are for your personal recovery or how prepared you are for a relapse, I can tell you it won't be easy to see the forest for the trees. You might have a distress kit full of all the right strategies and techniques for coping with your mental health disorder, but when the storm hits, it will still hurt, and you may very well find yourself face flat on the floor. What matters in such moments is the mindset you possess.

Believe it or not, your mindset will determine whether or not you recover and lead the life of your dreams. All the therapies, techniques, medication, and healing concepts in the world will be to no avail if you don't train your mindset to hold the right perspective. And if you think about it for a moment, you will realize that each time you've slipped or quit on yourself, it was because of the mindset you had. When something didn't go according to plan, you beat yourself up, engaged in unhealthy thinking and unhealthy behavior, which only made things worse. You then validated to yourself that you are, in fact, as inadequate, worthless, and as helpless as you feared. That is how you move further away from healing and deeper into misery and illness.

Now that you've read this book and increased your awareness, you can develop the kind of mindset that will lead you to that end objective of health and happiness and regain control if a setback happens. Here's how.

#1. Acknowledge that you've fallen off track.

Sometimes it's hard to recognize a setback, but if you have a healthy support system of people who care for you, they will point it out. Whether you realize it on your own or it comes from someone else, do not allow yourself that old impulsive behavior or victimhood and denial. Instead, I encourage you to be open to this new awareness. Remain receptive and objectively look at the facts being presented. Then move onto the next step.

#2. Tell yourself that it's okay.

Most research suggests that over 50% of people in recovery will experience a setback at some point in their journey even if they take every precaution. That tells me that it's normal and expected. So, in other words, you shouldn't waste your energy making a relapse or setback something horrible that should be avoided at all cost. Instead, you should be figuring out the most effective ways for you to manage a setback if and when it should arise. When it does, remind yourself that you are prepared for this.

#3. Accept full responsibility.

Aside from denial, the next worst thing you can do is blame the setback on something or someone. Sure, you were triggered, and that threw you off track but putting the blame on that trigger weakens you and steals your ability to take back control. That's why the next important thing to do is to assume responsibility for this experience. The choices you made inadvertently led you astray. This isn't to say you should now beat yourself up or justify what others did. Absolutely not! Even though others might have had a role in your difficulties, don't play the blame game. It will immobilize you and make your condition worse. What we want is to gain back control as soon as possible. So, we take full responsibility for our state and mental health and immediately start to ask high-quality empowering questions such as "what can I learn from this to make myself better?"

#4. Get help.

Trying to overcome a relapse or setback on your own is not advisable. The best and quickest way to get back on track is to call someone for support and ask for help. This can be your therapist, a counselor, or even a trusted friend. Immediately ask for some assistance even if the issue was small. Why? Because when a setback happens, the main focus should be getting back on track, and often our ego, sense of failure, or doubt can get in the way. So, we call reinforcement to help us rebuild that confidence and create a safe environment where the risk of further damage is mitigated. That will allow you the opportunity to regroup and motivate yourself again.

IT'S NOT THE END: WHAT TO REMEMBER

A question that I've been getting lately from friends is this.

"How do I get over this feeling of being a total failure after I have a relapse?"

It's a valid question and often challenging to give a simple answer because it all depends on why you have come to that conclusion in the first place. A story of a close friend comes to mind that might help you reconcile your concerns. He and I supported each other throughout our personal recovery, and we shared similar ambitions of recreating our lives. We both did tremendously well in the last few years, and he is now a thriving real estate agent working for one of the biggest companies in our city.

The last relapse almost broke him down again because he felt he had done everything in his power to get better and rebuild this new life, yet he still had a major relapse. I saw it as just a minor setback. His doctor assured him it was just a relapse. He felt different. It was catastrophic and a sign that he was never going to be anything but a complete failure. Of course, it wasn't any of those things.

What was happening here is that my friend was having a hard time coming to terms with what a mental health relapse means. So, it was up to us (his support structure) to remind him that recovery is never linear and that we all fluctuate from season to season. It's not the setback that matters but the mindset. So, I want to issue the same reassurance to you by reminding you of a few things.

First and foremost, you must avoid beating yourself up when you do have a relapse. That will only make things worse and disempower you. The second thing I want you to remember is that this too shall pass. You pulled yourself out of that pit of despair when you recovered before this setback, so this isn't an impossible mission. You can do this as many times as needed. What matters is that you've proven that you can function in life without the burden of illness that tortures you. Yes, it may take weeks or even months before you're up and running again in top form, but there's no reason for you to accept defeat. If you need to get another round of treatment, resume taking the medication prescribed by your therapist or whatever else, there's no shame in that. This is an on-going journey, and just as you will experience bumps in the road when taking care of your physical health, it's normal to have the same with your mental health.

CONCLUSION

Congratulations on sticking with this till the end. You've now learned what Cognitive Behavioral Therapy, Dialectical Behavioral Therapy, and Acceptance Commitment Therapy are, plus how they work. You also learned various ways of applying both therapy and self-care techniques to heal your mental health disorders. You should feel very proud for having the courage to take this first step. Taking a long and hard look at your destructive thought patterns and opening up your mind to new concepts that challenge your belief system and behavior is never easy. Your efforts will soon pay off.

As you continue applying what you learned, you will start to realize that people and situations don't drain or drag you down as much as they used to. Sure, you'll still face the trials and stresses of everyday life, but they won't seem like such a big deal; you'll find it easier with time to just shake things off and reset.

Best of all, you'll learn to stop taking everything so personally. You will detach from the need to value yourself based on other people's words and actions. What someone says or does or fails to do or say will no longer negatively impact your sense of worth.

All the techniques you've learned here (CBT, DBT, or ACT) will get easier as you practice them daily. Think of it like driving a car. Remember how awkward and complicated it felt the first time you sat behind the wheel? It didn't seem possible that one day you could automatically drive without giving it too much thought, and yet, if you've been doing it a while, you can probably attest to the fact that some days it feels like the car drives itself. Am I right? In just a matter of months, something as foreign as operating a machine to move from point A to B suddenly becomes second nature. But to get there, you had to invest time, money, and effort in driving lessons.

The same goes for CBT. As you make an investment in training your brain, you will reap the rewards for the rest of your life. Things that may seem hard now will become second nature, and the effort put in will be worth it when you look back from your newly rebuilt active and healthy lifestyle.

For years, therapy was considered something shameful, a taboo in some cultures. But thankfully, things have changed, and people realize taking care of your mental health is just as important and probably no different from signing up for that annual gym membership with a personal trainer. Learning how to control and regulate your emotions, stop irrational thoughts from ruling your mind, and eliminate detrimental habits from your routine is impactful to you and your entire community. The better you feel, the more you can make a difference in the world.

If you take only one thing away from this book, let it be that you now have more power over your mental health issue than you did before starting this book. You have a fully equipped distress kit with all kinds of wonderful techniques for mindfulness practices and hacks you can use to handle cognitive distortions and entire frameworks that can help with cognitive restructuring. The methods offered in CBT, DBT, and ACT can be used for the rest of your life with no harmful side effects, so don't just see this as a quick fix. Instead, approach these techniques as your allies on this recovery journey and turn them into lifelong friends that you can always lean on. Ultimately, this book has shown you that you can regain control and rebuild your life no matter how bad things have been. Realize this simple yet profound truth for yourself. The next step is to wear this new mindset of empowerment as much as you can from now on.

RESOURCES

.

Holland, K. (2020, September 3). Everything You Need to Know About Anxiety. Retrieved January 26, 2021, from https://www.healthline.com/health/anxiety

Higuera, V. (2020, February 11). Everything You Want to Know About Depression. Retrieved January 26, 2021, from https://www.healthline.com/health/depression

NIMH » Obsessive-Compulsive Disorder. (2021, January 28). Retrieved January 26, 2021, from https://www.nimh.nih.gov/health/topics/obsessive-compulsive-disorder-ocd/index.shtml

Why Cognitive Behavioral Therapy Is the Current Gold Standard of Psychotherapy. (2018). Retrieved January 26, 2021, from https://www.ncbi.nlm.nih.gov/pmc/articles/PMC5797481/

Psycom.net. (2019, October 21). Dialectical Behavior Therapy (DBT): Is it Right for You? Retrieved January 26, 2021, from https://www.psycom.net/what-is-dialectical-behavior-therapy/

Effectiveness of Acceptance and Commitment Therapy compared to CBT+: Preliminary results. (2018, October 1). Retrieved January 26, 2021, from https://www.sciencedirect.com/science/article/abs/pii/S0213616317301398

Effectiveness of Acceptance and Commitment Therapy on Anxiety and Depression of Razi Psychiatric Center Staff. (2018, February 15). Retrieved January 26, 2021, from https://www.ncbi.nlm.nih.gov/pmc/articles/PMC5839459/

Stanborough, R. M. J. (2020, February 4). How to Change Negative Thinking with Cognitive Restructuring. Retrieved January 26, 2021, from https://www.healthline.com/health/cognitive-restructuring

Time Management and Procrastination. (n.d.). Retrieved January 26, 2021, from https://caps.ucsc.edu/resources/time-management.html

GoodTherapy Editor Team. (n.d.). Mindfulness-Based Interventions. Retrieved January 26, 2021, from https://www.goodtherapy.org/learn-about-therapy/types/mindfulness-based-interventions

5 Simple ness Practices for Daily Life. (2018, December 13). Retrieved January 26, 2021, from https://www.mindful.org/take-a-mindful-moment-5-simple-practices-for-daily-life/

Skrzypińska, K. (2020, February 27). Does Spiritual Intelligence (SI) Exist? A Theoretical Investigation of a Tool Useful for Finding the Meaning of Life. Retrieved January 26, 2021, from https://link.

springer.com/article/10.1007/s10943-020-01005-8?error=
cookies_not_supported&code=799c0892-0deb-47eb-920f-
51f04abfedcc

M. (n.d.-a). Coping with Depression - HelpGuide.org. Retrieved
January 26, 2021, from https://www.helpguide.org/articles/
depression/coping-with-depression.htm

Neuropeak Pro. (2020, December 9). 10 Habits to Improve Your
Mood. Retrieved January 26, 2021, from https://www.neuropeakpro.
com/10-habits-to-improve-your-mood/

St.Cyr, J. L. (2019, May 22). Why Relaxation Techniques Don't Work
for Trauma & What to Do Instead. Retrieved January 26, 2021, from
https://healingwellcounseling.com/blog/trauma-survivors-why-
relaxation-techniques-dont-work-and-what-to-do-instead/

Setbacks in Mental Health Recovery Do Not Ruin Your Recovery |
HealthyPlace. (2017, February 19). Retrieved January 26, 2021, from
https://www.healthyplace.com/blogs/survivingmentalhealthstigma/
2017/02/setbacks-dont-ruin-mental-health-recovery

M. (n.d.). Finding a Therapist Who Can Help You Heal - Help-
Guide.org. Retrieved January 26, 2021, from https://www.helpguide.
org/articles/mental-health/finding-a-therapist-who-can-help-you-
heal.htm

B. (2016, August 5). How to Know if Therapy is Working. Retrieved
January 26, 2021, from https://beckinstitute.org/how-to-know-if-
therapy-is-working/

Community Mental Health Team for Older People, NSH Tayside. (n.d.). Staying Well – Relapse Prevention. Retrieved January 26, 2021, from https://www.nhstaysidecdn.scot.nhs.uk/NHSTaysideWeb/idcplg?IdcService=GET_SECURE_FILE&dDocName=PROD_233765&Rendition=web&RevisionSelectionMethod=LatestReleased&no

Flanigan, R. L. (2020, August 28). Depression & Relapse: Learning from the Setbacks. Retrieved January 26, 2021, from https://www.hopetocope.com/depression-relapse-learning-from-the-setback/

Overcoming Setbacks With Depression or Anxiety. (n.d.). Retrieved January 26, 2021, from https://www.premierhealth.com/your-health/articles/women-wisdom-wellness-/Overcoming-Setbacks-With-Depression-or-Anxiety/

Lightning Source UK Ltd.
Milton Keynes UK
UKHW010631190822
407545UK00001B/38